CALCULATIONS FOR CATERING STUDENTS

Other titles in this series:

Accounting in the Hotel and Catering Industry
Frank Wood and Peter Lightowlers

Purchasing, Costing and Finance in the Hotel and Catering Industry
Frank Wood and Peter Lightowlers

Basic Restaurant Theory and Practice
M. Anker and V.K. Batta

Advanced Restaurant Theory and Practice
(forthcoming)
M. Anker

Basic Science for Food Studies
V.L. Brownsell, C.J. Griffith and Eleri Jones

Applied Science for Food Studies
V.L. Brownsell, C.J. Griffith and Eleri Jones

Business Aspects of Catering
D. Ryan

H.G. DAVIES

CALCULATIONS FOR CATERING STUDENTS

Longman
Scientific &
Technical

Longman Scientific & Technical
Longman Group UK Limited,
Longman House, Burnt Mill, Harlow,
Essex CM20 2JE, England
and Associated Companies throughout the world.

First published 1989

British Library Cataloguing in Publication Data

Davies, H.G. (Henri Gwyn)
Calculations for catering students.
1. Mathematics — For catering students
I. Title
510′.24642

ISBN 0-582-41265-X

Produced by Longman Singapore Publishers (Pte) Ltd.
Printed in Singapore

Cyflwynaf y llyfr hwn i
Ceris,
Catrin, Sion a Leisa

PREFACE

The aim of this book is to provide a mathematical basis for the range of calculations that are commonly used in the Catering Industry. The material in the book is not written with any particular syllabus in mind, but is based on the examination syllabuses of CGLI, BTEC and HCIMA.

Worked examples are included in the text where relevant, and exercises are included within chapter. At the end of each chapter a Miscellaneous Exercise is included, such exercises are included in the book in an attempt to arrive at some basic standard of presentation.

For many professional colleagues I am extremely indebted for advice and encouragement over many years:

1 To the staff of the Catering Department of Carmarthenshire College of Technology and Art, where I was Head of Department for many years.
2 To Mr John Griffiths, Dean of the Faculty of Catering and Food Technology, for the many hours that I prevailed upon him.
3 To M Jean Bouchard and staff of Les Sorbet, in Nour Moutier, for the opportunity to exchange ideas.
4 To Miss Eirlys Peck for typing the manuscript.

CONTENTS

ACKNOWLEDGEMENTS

To the following Colleges I am indebted for permission to include exercises from their examination papers.

Neath	(Neath College)
Nelson and Colne	
Salisbury	(Salisbury College of Technology)
Blackpool and Fylde	(Blackpool and Fylde College of Further and Higher Eduction)
GLOSCAT	(Gloucester College of Art and Technology)
Oxford	(Oxford College for Further Education)
HCIMA	(Hotel Catering and Institutional Management Association)

1
FRACTIONS

1.1 RULES OF PRECEDENCE

In working with numbers four basic operations are used, namely addition, subtraction, multiplication and division, such as:

$$3 + 4$$
$$6 - 2$$
$$7 \times 3$$
$$18 \div 6$$

When a calculation contains more than two of these basic operations then a set of rules is required to state clearly the order in which the operations must be carried out. For example, in the calculation:

$$17 - 2 \times 3$$

we must agree on which of the two operations ($-$), ($\times$) is to be carried out first, because changing the order gives different answers, as seen below:

Carrying out the subtraction first gives:
$(17 - 2) \times 3 = 15 \times 3 = 45$

Carrying out the multiplication first gives:
$17 - (2 \times 3) = 17 - 6 = 11$

The order in which operations are to be carried out is as follows:

(i) Anything in a pair of brackets
(ii) Multiplication and division
(iii) Addition and subtraction.

This list constitutes the rules of precedence. Remembering these rules of precedence is extremely important when using the electronic calculator, as seen in Chapter 3.

The application of the rules of precedence is shown in the following examples

EXAMPLE 1
Evaluate $3 + 7 \times 6 - 4 \div 2$

In accordance with the rules of precedence, the multiplication and division are carried out first, to give:

$$\begin{aligned} 3 + (7 \times 6) - (4 \div 2) &= 3 + 42 - 2 \\ &= 45 - 2 \\ &= 43 \end{aligned}$$

Note that the brackets are only introduced for convenience.

EXAMPLE 2
Work out $(7 - 3) \times 3 \div 4$

In this calculation the inside of the bracket must be evaluated first, that is:

$$\begin{aligned} (7 - 3) \times 3 \div 4 &= 4 \times 3 \div 4 \\ &= 12 \div 4 \\ &= 3 \end{aligned}$$

EXAMPLE 3
Calculate $(4 \times 2 - 3) + 15 \div 5$

The inside of the bracket must be evaluated first and within the bracket the multiplication takes precedence.

$$\begin{aligned} (4 \times 2 - 3) + 15 \div 5 &= (8 - 3) + 15 \div 5 \\ &= 5 + 15 \div 5 \\ &= 5 + 3 \\ &= 8 \end{aligned}$$

EXERCISES 1.1

Using the rules of precedence, evaluate the following:

1. $14 + 7 - 3 \times 5$
2. $7 \times 2 - 4 \times 3$
3. $15 \times 2 \div 3 - 6$
4. $7 - 4 \times 2 + 6$
5. $13 - 14 \div 7 + 6$
6. $(6 + 3) - 4 \times 2$
7. $(8 + 4 \times 2) - 10$
8. $6 - 3 \times 2 + 7$
9. $(6 - 3) \times 2 + 7$
10. $(6 - 3) \times (2 + 7)$

1.2 THREE LAWS OF ARITHMETIC

(a) *COMMUTATIVE LAW*

It is well known that when two numbers are added or multiplied together, it does not matter in which order the numbers are written down. For example:

$$6 + 7 = 7 + 6$$
and $$8 \times 5 = 5 \times 8$$

These two examples illustrate the commutative law.

(b) *ASSOCIATIVE LAW*

This law applies to the addition or multiplication of more than two numbers and states that the order in which the operation is carried out does not change the answer. Note that we are now dealing with *all* multiplication or *all* addition, and not the mixed operation of section 1.1. For example:

$$4 + (6 + 8) = (4 + 6) + 8$$
and $$2 \times (3 \times 7) = (2 \times 3) \times 7$$

(c) *DISTRIBUTIVE LAW*

The distributive law applies to the multiplication of the sums or differences of numbers inside a bracket by a number outside, that is:

$$7 \times (4 + 3)$$

The law states that this operation can be written as:

$$7 \times 4 + 7 \times 3$$

This can be verified quite easily:

$$7 \times (4 + 3) = 7 \times 7 = 49$$
$$7 \times 4 + 7 \times 3 = 28 + 21 = 49$$

1.3 FRACTIONS

(a) *THE SIMPLIFICATION OF FRACTIONS*

Consider a gateau cut into 8 equal pieces as shown in Fig. 1.1. If the two shaded pieces are removed, then 2 parts out of 8 parts are removed. This

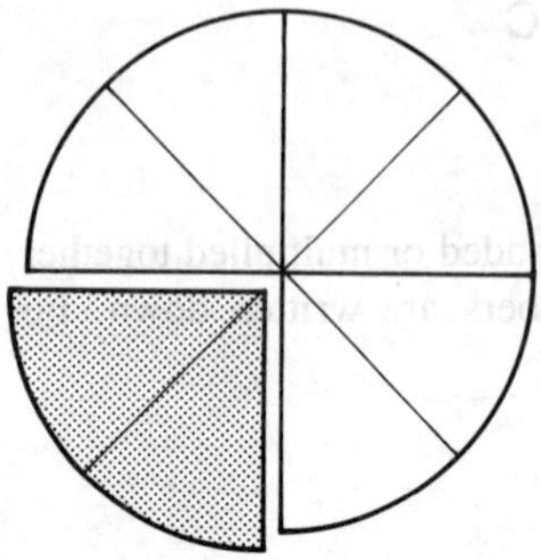

FIG. 1.1 A gateau divided into 8 pieces

is written as $\frac{2}{8}$, and is called a **fraction**. It is the fraction of the whole gateau that is removed. The number 2 above the line is called the **numerator** and below the line the number 8 is the **denominator**.

By reference to the diagram, it is seen that the two pieces represent $\frac{1}{4}$ (one-quarter) of the gateau. Therefore:

$$\frac{2}{8} = \frac{1}{4}$$

This shows an important feature of fractions. They may be simplified in many cases by dividing the numerator and denominator by the same number. In this example, 2 divided out top and bottom. This process is illustrated again in Example 4.

EXAMPLE 4

A tray of apple tart is divided into 60 pieces as shown in Fig. 1.2. At a first sitting of lunch, 18 pieces are served. Using the diagram find the fraction of the tray served.

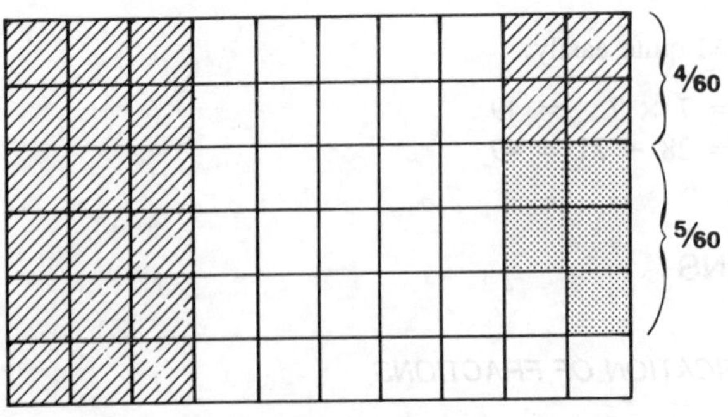

FIG. 1.2 A slab of apple tart divided into 60 pieces

The pieces served during the first sitting are shown by the shaded portion on the left-hand side of the diagram.

The fraction served is $\frac{18}{60}$.

It is seen however, that this fraction is also 3 rows out of 10 rows, that is $\frac{3}{10}$.

$$\therefore \frac{18}{60} = \frac{3}{10}$$

This can be obtained by dividing out top and bottom, by 6. Such a process is called 'cancelling':

$$\frac{\overset{3}{18}}{\underset{10}{60}} = \frac{3}{10}$$

EXAMPLE 5

Reduce the following fractions to the simplest form:

(a) $\frac{10}{25}$; (b) $\frac{8}{12}$.

(a) 5 will divide into top and bottom:

$$\frac{\overset{2}{10}}{\underset{5}{25}} = \frac{2}{5}$$

(b) 4 will divide into top and bottom:

$$\frac{\overset{2}{8}}{\underset{3}{12}} = \frac{2}{3}$$

(b) *TYPES OF FRACTIONS*

A fraction such as $\frac{2}{5}$ where the top (numerator) is smaller than the bottom (denominator) is called a **proper** fraction.

A fraction such as $\frac{7}{5}$ is top-heavy and is called an **improper** fraction.

A fraction such as $1\frac{2}{5}$ is a **mixed** fraction.

Note: Improper and mixed fractions are different forms of the same number as shown below:

$$1\tfrac{2}{5} = 1 + \tfrac{2}{5} = \tfrac{5}{5} + \tfrac{2}{5} = \tfrac{7}{5}$$

$$2\tfrac{3}{7} = 2 + \tfrac{3}{7} = \tfrac{14}{7} + \tfrac{3}{7} = \tfrac{17}{7}$$

In the first mixed fraction we find how many fifths there are in 1, which is 5, i.e. $\frac{5}{5}$. Similarly, in the second fraction, how many sevenths there are in 2, which is 14, i.e. $\frac{14}{7}$.

To convert an improper fraction back into a mixed fraction, find how many times the bottom will divide into the top. In $\frac{17}{7}$, 7 will divide into 17 twice, leaving 3 over, that is:

$$\frac{17}{7} = 2 + \frac{3}{7} = 2\frac{3}{7}$$

Again

$$\frac{13}{4} = 3 + \frac{1}{4} = 3\frac{1}{4}$$

EXERCISES 1.2

1. Simplify the following fractions:

(a) $\frac{16}{20}$

(b) $\frac{9}{12}$

(c) $\frac{30}{50}$

(d) $\frac{18}{54}$

(e) $\frac{15}{65}$

2. Express the following as mixed fractions:

(a) $\frac{17}{2}$

(b) $\frac{9}{7}$

(c) $\frac{112}{9}$

(d) $\frac{14}{13}$

3. Express the following as improper fractions:

(a) $1\frac{5}{7}$

(b) $4\frac{1}{3}$

(c) $2\frac{2}{5}$

(d) $6\frac{1}{7}$

(c) *ADDITION OF FRACTIONS*

Consider the addition $\frac{4}{60} + \frac{5}{60}$. From Fig. 1.2 this is seen to be $\frac{9}{60}$, shown by the shaded areas on the right-hand side of the diagram. When the two fractions have the same denominator, the addition becomes quite easy.

For addition such as $\frac{3}{5} + \frac{7}{10}$ therefore, the first requirement is to express both fractions in the same denominator. Therefore:

$$\frac{3}{5} + \frac{7}{10} = \frac{6}{10} + \frac{7}{10}$$

$$= \frac{13}{10}$$

$$= 1\frac{3}{10}$$

The pattern for carrying out addition may be written as follows:

(i) Find the smallest number into which all the denominators will divide exactly. This is called the **lowest common multiple** (LCM).
(ii) Express all fractions with this LCM as denominators.

Example 6 shows the procedure.

EXAMPLE 6

Add $\frac{1}{3} + \frac{3}{4} + \frac{1}{2}$ ____________

The LCM is 12. Expressing the three fractions with the denominator of 12, gives:

$$\frac{1}{3} = \frac{*}{12} = \frac{4}{12}$$

$$\frac{3}{4} = \frac{*}{12} = \frac{9}{12}$$

$$\frac{1}{2} = \frac{*}{12} = \frac{6}{12}$$

$$\text{Add: } \frac{19}{12} = 1\frac{7}{12}$$

EXAMPLE 7

Add $\frac{1}{5} + \frac{5}{6} + \frac{1}{2}$ ____________

The LCM is 30. Expressing each fraction with 30 as denominator gives:

$$\frac{1}{5} = \frac{*}{30} = \frac{6}{30}$$

$$\frac{5}{6} = \frac{*}{30} = \frac{25}{30}$$

$$\frac{1}{2} = \frac{*}{30} = \frac{15}{30}$$

$$\text{Add: } \frac{46}{30} = 1\frac{16}{30}$$

$$= 1\frac{8}{15}$$

If the fractions to be added are mixed fractions the whole numbers are evaluated first, and then the fractions as shown already.

EXAMPLE 8

Add $2\frac{1}{4} + 1\frac{2}{5} + 1\frac{1}{6}$

(i) Adding the whole numbers first: $4 + \frac{1}{4} + \frac{2}{5} + \frac{1}{6}$
(ii) LCM is 60
(iii) The sum becomes:

$$4 + \frac{*}{60} + \frac{*}{60} + \frac{*}{60}$$

$$= 4 + \frac{15}{60} + \frac{24}{60} + \frac{10}{60}$$

$$= 4 + \frac{49}{60}$$

$$= 4\frac{49}{60}$$

(d) *SUBTRACTION OF FRACTIONS*

The same process applies to subtraction, as shown in Example 9.

EXAMPLE 9

Evaluate $4\frac{3}{4} - 1\frac{1}{6}$

(i) Subtract the whole numbers first: $3 + \frac{3}{4} - \frac{1}{6}$
(ii) LCM: 12
(iii) The subtraction becomes:

$$3 + \frac{*}{12} - \frac{*}{12}$$

$$= 3 + \frac{9}{12} - \frac{2}{12}$$

$$= 3 + \frac{7}{12}$$

$$= 3\frac{7}{12}$$

It often happens that the second proper fraction in such a subtraction is greater than the first, as in Example 10.

EXAMPLE 10

Subtract $5\frac{2}{3} - 2\frac{4}{5}$ ________________

Proceeding as before, we have:

(i) Subtract the whole numbers first: $3 + \frac{2}{3} - \frac{4}{5}$
(ii) LCM: 15
(iii) Subtraction becomes:

$$3 + \frac{*}{15} - \frac{*}{15}$$

$$= 3 + \frac{10}{15} - \frac{12}{15}$$

The 12 cannot be subtracted from 10. Therefore, one of the whole numbers is taken and used as $\frac{15}{15}$:

i.e.
$$3 + \frac{10}{15} - \frac{12}{15}$$

$$= \overbrace{2 + \frac{15}{15}} + \frac{10}{15} - \frac{12}{15}$$

$$= 2 + \frac{25}{15} - \frac{12}{15}$$

$$= 2\frac{13}{15}$$

A mixture of addition and subtraction can be carried out in a similar manner, as shown in Example 11.

EXAMPLE 11

Evaluate $4\frac{2}{5} + 1\frac{1}{4} - 2\frac{9}{10}$ ________________

(i) Evaluate the whole number first: $3 + \frac{2}{5} + \frac{1}{4} - \frac{9}{10}$
(ii) LCM: 20
(iii) The calculation becomes:

$$3 + \frac{8}{20} + \frac{5}{20} - \frac{18}{20}$$

$$= 3 + \frac{13}{20} - \frac{18}{20}$$

(iv) 18 cannot be subtracted from 13. One of the whole numbers is converted into $\frac{20}{20}$, to give:

$$2 + \tfrac{20}{20} + \tfrac{13}{20} - \tfrac{18}{20}$$

$$= 2 + \tfrac{15}{20}$$

$$= 2\tfrac{15}{20}$$

$$= 2\tfrac{3}{4} \quad \text{on simplifying}$$

EXERCISES 1.3

1. Carry out the following additions:

(a) $\frac{3}{4} + \frac{1}{5}$

(b) $\frac{7}{8} + \frac{2}{3}$

(c) $\frac{3}{5} + \frac{3}{10} + \frac{1}{4}$

(d) $3\frac{4}{7} + 1\frac{1}{4} + 4\frac{1}{2}$

(e) $2\frac{2}{3} + 1\frac{7}{9} + 1\frac{1}{4}$

(f) $2\frac{3}{8} + 1\frac{4}{5} + 5\frac{7}{10}$

2. Subtract the following:

(a) $\frac{4}{7} - \frac{1}{3}$

(b) $4\frac{7}{9} - 1\frac{3}{4}$

(c) $3\frac{3}{4} - 1\frac{2}{5}$

(d) $3\frac{7}{12} - 1\frac{3}{4}$

(e) $4\frac{3}{8} - 1\frac{2}{3}$

(f) $5\frac{1}{7} - 2\frac{2}{3}$

3. Evaluate the following:

(a) $4\frac{1}{3} + 2\frac{1}{4} - 1\frac{5}{6}$

(b) $3\frac{5}{8} - 1\frac{1}{3} - \frac{5}{12}$

(c) $6\frac{2}{3} - 1\frac{4}{5} - 1\frac{7}{10}$

(d) $5\frac{1}{7} + 3\frac{3}{4} - 4\frac{1}{2}$

(e) $5\frac{4}{7} + 1\frac{1}{3} - 4\frac{1}{2}$

(f) $2\frac{5}{6} - 1\frac{2}{5} + 3\frac{2}{3}$

(e) *MULTIPLICATION OF FRACTIONS*

Consider the following exercise: A batch of 240 *petits fours* are prepared for the restaurant by the TEC Diploma class. On the first day $\frac{3}{8}$ of the batch are used. How many *petits fours* were actually used?

The fraction $\frac{3}{8}$ means that 3 out of 8 were used up. The number of groups of 8 in the whole batch of *petits fours* is

$$240 \div 8 = 30$$

From each of these 30 groups, 3 are used up on the first day, that is,

$$30 \times 3 = 90$$

Thus, the actual number used = 90

The process is seen to be that of dividing by the denominator and multiplying by the numerator, that is:

$$\begin{aligned} \tfrac{3}{8} \text{ of } 240 &= \frac{3 \times 240}{8} \\ &= 3 \times 30 \\ &= 90 \end{aligned}$$

EXAMPLE 12

A restaurant served 900 lunches in one week. Of these, $\frac{2}{5}$ were served on Friday. How many meals were actually served on this day?

$$\begin{aligned} \text{Number of meals served} &= \tfrac{2}{5} \text{ of } 900 \\ &= \frac{2 \times 900}{5} \\ &= 2 \times 180 \\ &= 360 \end{aligned}$$

The above process can be extended to the multiplication of two fractions, such as:

$\frac{3}{4} \times \frac{5}{7}$

Now the two numerators are multiplied together and also the denominator, to give:

$$\frac{3 \times 5}{4 \times 7} = \frac{15}{28}$$

As already seen in Example 12, **of** is used in the same way as multiplication. Therefore,

$$\frac{4}{5} \text{ of } \frac{7}{9} = \frac{4}{5} \times \frac{7}{9}$$
$$= \frac{28}{45}$$

In carrying out any multiplication of fractions, simplification should be carried out by 'cancelling' as explained in Example 4.

Therefore, $\frac{\overset{1}{\not{3}}}{5} \times \frac{4}{\underset{3}{\not{9}}} = \frac{1 \times 4}{5 \times 3}$

$$= \frac{4}{15}$$

EXAMPLE 13

Evaluate $\frac{8}{15} \times \frac{25}{28}$ ____________________

Cancelling out with 4 and 5 gives:

$$\frac{\overset{2}{\not{8}}}{\underset{3}{\not{15}}} \times \frac{\overset{5}{\not{25}}}{\underset{7}{\not{28}}} = \frac{2 \times 5}{3 \times 7}$$

$$= \frac{10}{21}$$

In multiplying out mixed fractions it is first necessary to change such fractions into top-heavy fractions and then to proceed as in Example 13. This is shown in Example 14.

EXAMPLE 14

Evaluate $4\frac{1}{5} \times 2\frac{1}{7}$ ____________________

Changing to top-heavy fractions and cancelling with 5 and 7 gives:

$$\frac{21}{5} \times \frac{15}{7}$$

$$= \frac{{}^{3}\cancel{21}}{{}_{1}\cancel{5}} \times \frac{\cancel{15}^{3}}{\cancel{7}_{1}} = \frac{3 \times 3}{1 \times 1}$$

$$= \frac{9}{1}$$

$$= 9$$

(f) *DIVISION OF FRACTIONS*

The method of dividing fractions can be explained as follows:

8 apples divided into pairs gives 4 groups
i.e. $8 \div 2 = 4$
8 apples divided into halves gives 16
i.e. $8 \div \frac{1}{2} = 16$
8 apples divided into quarters gives 32
i.e. $8 \div \frac{1}{4} = 32$
8 apples divided into eighths gives 64
i.e. $8 \div \frac{1}{8} = 64$

All these simple deductions are examples of division by fractions. The results can be obtained by the following division rule:

RULE

> The fraction on the right of the division sign is inverted and the division changed to a multiplication.

The above examples using this rule can be seen to be:

$8 \div \frac{2}{1} = \frac{8}{1} \times \frac{1}{2} = 4$

$8 \div \frac{1}{2} = \frac{8}{1} \times \frac{2}{1} = 16$

$8 \div \frac{1}{4} = \frac{8}{1} \times \frac{4}{1} = 32$

$8 \div \frac{1}{8} = \frac{8}{1} \times \frac{8}{1} = 64$

As with multiplication, mixed fractions are first converted into top-heavy fractions. Therefore:

$$
\begin{aligned}
2\tfrac{1}{4} \div 1\tfrac{1}{2} &= \tfrac{9}{4} \div \tfrac{3}{2} \\
&= \frac{\cancel{9}^{\,3}}{\cancel{4}_{\,2}} \times \frac{\cancel{2}^{\,1}}{\cancel{3}_{\,1}} \\
&= \tfrac{3}{2} \\
&= 1\tfrac{1}{2}
\end{aligned}
$$

EXAMPLE 15

Calculate $4\tfrac{4}{5} \div 1\tfrac{5}{11}$

$$
\begin{aligned}
4\tfrac{4}{5} \div 1\tfrac{5}{11} &= \tfrac{24}{5} \div \tfrac{16}{11} \\
&= \frac{\cancel{24}^{\,3}}{5} \times \frac{11}{\cancel{16}_{\,2}} \\
&= \tfrac{33}{10} \\
&= 3\tfrac{3}{10}
\end{aligned}
$$

EXERCISES 1.4

Evaluate the following:

1. $6\tfrac{1}{4} \times 2\tfrac{2}{5}$

2. $1\tfrac{1}{9} \times 1\tfrac{4}{35}$

3. $2\tfrac{2}{21} \times 6\tfrac{4}{11}$

4. $5\tfrac{5}{8} \times 1\tfrac{17}{27}$

5. $1\tfrac{1}{15} \div 1\tfrac{7}{25}$

6. $4\tfrac{4}{9} \div 2\tfrac{2}{15}$

7. $2\tfrac{2}{13} \div 1\tfrac{9}{26}$

8. $2\tfrac{2}{5} \div 1\tfrac{1}{3}$

9. $3\frac{1}{3} \times 1\frac{2}{5} \times 2\frac{5}{14}$

10. $1\frac{4}{11} \times 2\frac{5}{14} \times 2\frac{1}{10}$

11. $1\frac{3}{5} \div 2\frac{2}{21} \div 2\frac{6}{11}$

12. $1\frac{3}{8} \times 3\frac{3}{7} \div 4\frac{5}{7}$

(g) *MIXED OPERATIONS INVOLVING +, −, ×, ÷*

The rules of precedence stated in section 1.1 also apply to the fractions, that is:

(i) the inside of any brackets is worked out first,
(ii) × and ÷ are carried out before + and −.

EXAMPLE 16
Evaluate

(a) $1\frac{3}{4} + 2\frac{2}{5} \times 3\frac{1}{8}$

(b) $(1\frac{2}{3} + 1\frac{1}{4}) \times 1\frac{1}{14}$

(a) The multiplication is carried out first:

$$
\begin{aligned}
1\tfrac{3}{4} + 2\tfrac{2}{5} \times 3\tfrac{1}{8} &= 1\tfrac{3}{4} + \frac{\overset{3}{\cancel{12}}}{\underset{1}{\cancel{5}}} \times \frac{\overset{5}{\cancel{25}}}{\underset{2}{\cancel{8}}} \\
&= 1\tfrac{3}{4} + \tfrac{15}{2} \\
&= 1\tfrac{3}{4} + 7\tfrac{1}{2} \\
&= 8 + \frac{3 + 2}{4} \\
&= 8 + \tfrac{5}{4} \\
&= 8 + 1\tfrac{1}{4} \\
&= 9\tfrac{1}{4}
\end{aligned}
$$

(b) The bracket is evaluated first:

$$(1\tfrac{2}{3} + 1\tfrac{1}{4}) \times 1\tfrac{1}{14} = \left(2 + \frac{8 + 3}{12}\right) \times 1\tfrac{1}{14}$$

$$= 2\tfrac{11}{12} \times 1\tfrac{1}{14}$$

$$= \frac{\overset{5}{\cancel{35}}}{\underset{4}{\cancel{12}}} \times \frac{\overset{5}{\cancel{15}}}{\underset{2}{\cancel{14}}}$$

$$= \tfrac{25}{8}$$

$$= 3\tfrac{1}{8}$$

EXERCISES 1.5

Work out the following:

1. $(1\frac{3}{4} + 1\frac{2}{5}) \times 1\frac{1}{9}$

2. $1\frac{3}{4} + 1\frac{2}{5} \times 1\frac{1}{9}$

3. $(2\frac{2}{3} - 1\frac{1}{4}) \div 6\frac{3}{8}$

4. $2\frac{2}{3} - 1\frac{1}{4} \div 1\frac{3}{7}$

5. $2\frac{1}{7} + 1\frac{1}{5} \times 1\frac{3}{7} - 1\frac{3}{14}$

6. $(2\frac{1}{7} - 1\frac{1}{3}) \div (1\frac{3}{7} - 1\frac{3}{14})$

7. $4\frac{1}{5} \div 3\frac{1}{9} + 1\frac{1}{5} \times 2\frac{1}{4}$

8. $4\frac{1}{5} \div (1\frac{1}{9} + 1\frac{1}{5}) \times 7\frac{3}{7}$

MISCELLANEOUS EXERCISES 1

1. Write each of the following as a single fraction:

(a) $\frac{1}{8} \times \frac{4}{5} \div \frac{3}{10}$

(b) $(\frac{2}{5} + \frac{1}{4}) \div \frac{4}{5}$

(c) $\frac{2}{5} + \frac{1}{4} \times \frac{4}{5}$

(d) $(\frac{1}{4} + \frac{1}{3}) \div (1 - \frac{3}{8} - \frac{1}{2})$

(e) $\dfrac{2\frac{2}{3} + 1\frac{1}{5}}{5\frac{4}{5}}$

(f) $2\frac{1}{2} + 6\frac{1}{4} \times 4 - 16(\frac{2}{3} + \frac{1}{4})$

(GLOSCAT)

2. A hotel needed to re-carpet three rooms. The areas of the three rooms were $24\frac{1}{5}$ m^2, $18\frac{3}{4}$ m^2, $19\frac{7}{10}$ m^2. Find the total area of the three rooms in order to estimate the amount of carpet required.

3. For the preparation of melba toast for the restaurant over three days, the number of loaves used were $3\frac{1}{2}$, $2\frac{1}{4}$, $1\frac{5}{6}$. How many loaves were used in total?

4. The storekeeper issued potatoes to the kitchen. The chef received $4\frac{2}{5}$ kg and a further $6\frac{1}{10}$ kg; but later returned $3\frac{1}{3}$ kg. What was the actual amount used in the kitchen that day?

5. A kitchen has three electric stoves. On a particular afternoon the units used were $7\frac{1}{4}$, $8\frac{1}{10}$ and $3\frac{1}{5}$ units, respectively. What was the total number of units used?

   The following afternoon, the units used were $6\frac{1}{3}$, $4\frac{1}{4}$, $4\frac{5}{6}$. Find the total used again and the difference between the totals of the two afternoons.

6. A carpet in a hotel corridor measured $1\frac{2}{3} \times 2\frac{3}{5}$ m. Find its area.

   It was laid in a corridor $3\frac{1}{5}$ m $\times$ $2\frac{1}{2}$ m. Find the area of the surround, not covered by carpet.

7. $4\frac{3}{4}$ cases of fruit were received for the restaurant at £$2\frac{2}{5}$ per case. How much was the total bill for the fruit?

8. A large turkey needs to be cooked for $\frac{7}{9}$ hour for each pound weight (lb). If the turkey weighed $20\frac{1}{4}$ lb, for how long should the turkey be cooked?

9. In order to make tablecloths for the restaurant, strips of linen were required $1\frac{5}{6}$ m wide. How many strips could be obtained from a cloth $16\frac{1}{2}$ m wide?

10. A restaurant used $1\frac{3}{5}$ packets of paper serviettes each lunchtime. In stock in the store rooms were $20\frac{4}{5}$ packets. How many days' supply of serviettes were there?

11. (a) Work out $8\frac{3}{10} + 2\frac{1}{4} \times 3\frac{1}{5}$

(b) Two rooms in a hotel are to be carpeted. One room has an area of $8\frac{3}{10}$ m^2. The other room measures $2\frac{1}{4}$ m by $3\frac{1}{5}$ m. Using part (a), find the total area of the two rooms.

12. A kitchen assistant was sent to buy packets of serviettes. She was given petty cash to the value of £$9\frac{7}{10}$. The packets cost £$\frac{3}{5}$ each. How many whole packets could she buy?

13. A new barman trainee opened three bottles of whisky in error. At the end of the day, the bottles had left in them $\frac{11}{32}$, $\frac{1}{4}$, $\frac{5}{16}$. Find the total amount of spirit left in the bottles.

2

DECIMALS

2.1 THE MEANING OF A DECIMAL NUMBER AND CONVERSION FROM AND TO FRACTIONS

Consider the grid shown in Fig. 2.1(a). The shaded part is composed of 3 out of the 10 strips, that is, $\frac{3}{10}$ of the whole square. This $\frac{3}{10}$ is written as 0.3, which is the decimal number representing 3 out of 10.

Similarly, 0.6 represents 6 part out of 10 parts.

We can now extend this to work out the meaning of the decimal number 0.34. If we look at Fig. 2.1(b) the square is divided into 100 small squares. The shaded part now consists of 34 small squares, that is 34 squares from 100 squares. This shaded part represents $0.34 = \frac{34}{100}$. We see from Fig. 2.1(b) that:

$$\begin{aligned} 0.34 &= 3 \text{ strips and } 4 \text{ small squares} \\ &= \tfrac{3}{10} + \tfrac{4}{100} \end{aligned}$$

Hence, we see that 0.34 lies between 0.3 and 0.4.

This idea can be extended to 3 decimal places, such as 0.347. This number is $\frac{347}{1000}$, and it is seen to be

$$0.387 = \tfrac{3}{10} + \tfrac{4}{100} + \tfrac{7}{1000}$$

(a) *CONVERSION OF DECIMALS TO FRACTIONS*

Using the above discusson, we can see that:

$$0.6 = \frac{\overset{3}{\cancel{6}}}{\underset{5}{\cancel{10}}} = \frac{3}{5}$$

$$0.75 = \frac{75}{100} = \frac{3}{4}$$

$$0.513 = \frac{513}{1000}$$

(b) *CONVERSION OF FRACTIONS TO DECIMALS*

The conversion of decimals to fractions is easily carried out as already explained. Now we show how it is possible to convert fractions to decimals.

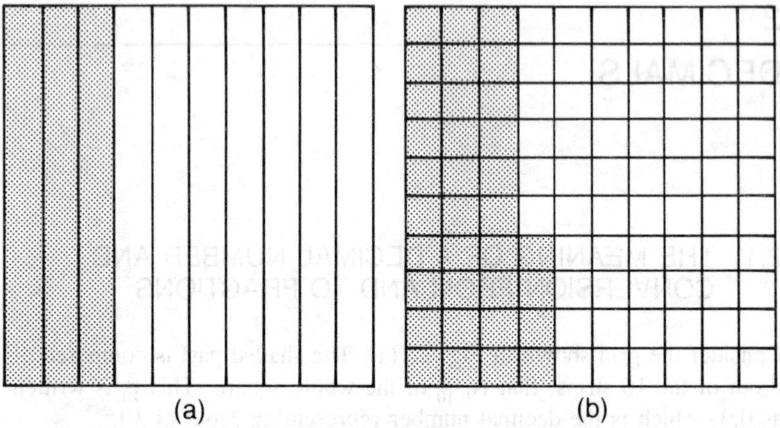

FIG. 2.1 (a) Shaded area $\frac{3}{10}$ of the whole (b) Shaded area $\frac{34}{100}$ of the whole square

If the fractions are tenths or hundredths, then the conversion is easy, that is:

$\frac{4}{10} = 0.4$

$\frac{31}{100} = 0.31$

Other fractions may be converted into decimals as follows:

(i) Insert the decimal point in the top number and add 0s.
(ii) Divide the bottom number into the top number.

EXAMPLE 1

Convert: (a) $\frac{3}{4}$, (b) $\frac{1}{8}$ to decimals.

(a) $\frac{3}{4}$

(i) Write the numerator as 3.00
(ii) Divide by 4

$$4 \,\underline{|\ 3.00}$$
$$0.75$$

i.e. $\frac{3}{4} = 0.75$

(b) $\frac{1}{8}$

(i) Write the numerator as 1.000
(ii) Divide by 8

$$8 \overline{)\,1.000}$$
$$0.125$$

i.e. $\frac{1}{8} = 0.125$

EXERCISES 2.1

1. Convert the following decimals to fractions:

(a) 0.5 (b) 0.6
(c) 0.45 (d) 0.64
(e) 0.72 (f) 0.96
(g) 0.725 (h) 1.5
(i) 2.25 (j) 3.22

Write the numbers out in decreasing size, starting with 3.22.

2. Convert the following fractions to decimals:

(a) $\frac{4}{10}$ (b) $\frac{34}{100}$

(c) $\frac{16}{100}$ (d) $\frac{4}{100}$

(e) $\frac{31}{1000}$ (f) $\frac{6}{1000}$

(g) $\frac{1}{4}$ (h) $\frac{1}{5}$

(i) $\frac{4}{5}$ (j) $\frac{7}{8}$

(k) $\frac{5}{8}$

After converting to decimals, write down the fractions in decreasing size.

2.2 MONEY AS DECIMAL NUMBERS

In the British currency,

$$£1 = 100p$$

Therefore, the £ is divided into a 100 parts. A sum of 36 parts of a £1 is:

$$36p = £\,\frac{36}{100}$$
$$= £0.36$$

Similarly, £4 and 47p = £4.47

Thus, it is seen that British currency reduces very nicely to decimal numbers with two decimal places.

2.3 RECURRING AND NON-TERMINATING DECIMAL NUMBERS

Consider the fraction $\frac{1}{3}$. Converting it into decimals gives:

$\frac{1}{3} = 0.33333\ldots$

The same number 3 carries on indefinitely. Such a decimal number is a **recurring** one. Other examples are:

$\frac{2}{3} = 0.66666\ldots$

$\frac{1}{9} = 0.11111\ldots$

$\frac{5}{9} = 0.55555\ldots$

Other decimal numbers exist, which are **non-terminating**. For example:

$\frac{1}{7} = 0.14285\ldots$

is a non-terminating decimal, that is, the fraction does not divide out exactly. Again,

$\frac{1}{13} = 0.0769\ldots$

2.4 REDUCTION TO A GIVEN NUMBER OF PLACES

A recurring or a non-terminating decimal goes on indefinitely. In order to use such numbers we must limit their length to a reasonable number of places. For example:

$\frac{2}{7} = 0.2857\downarrow1\ldots$

If we wish to limit the number to 4 places of decimals, shown by the position of the arrow, it becomes:

0.2857

We say that the number is **corrected** or **rounded off** to 4 decimal places.

If we wish to limit the same number to 3 decimal places, a small difficulty arises. The number as written is nearer to 0.286 than it is to 0.285, as shown in Fig. 2.2. The number corrected to 3 decimal places is therefore 0.286.

The rule for carrying out the rounding-off to a given number of places is obtained in BS 1957.

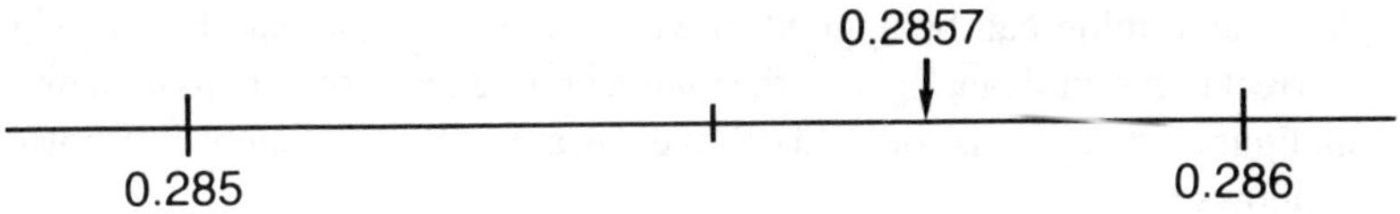

FIG. 2.2 The position of 0.2857

RULE

(i) If the digit to be removed is less than 5, the digit immediately preceding it is unchanged.
(ii) If the digit to be removed is greater than 5, the digit immediately preceding it is increased by 1.
(iii) If the digit to be removed is equal to 5 then the digit immediately preceding it is increased by 1, only if it is odd.

The rule is illustrated in Example 2.

EXAMPLE 2

Round off the following to 2 places of decimals:

(a) 3.214 (b) 2.319
(c) 2.455 (d) 6.325

(a) Using rule (i): 3.21
(b) Using rule (ii): 2.32
(c) Using rule (iii) the digit to be removed is 5. The preceding digit is odd and it is thus increased by 1:
i.e. 2.46
(d) As in (c), but the digit immediately preceding it is even and it is then unchanged,
i.e. 6.32

2.5 SIGNIFICANT FIGURES

Instead of rounding off a number to a number of decimal places it can be corrected to a number of **significant figures**. For example:

16.317

contains 5 figures, counting from the left-hand side. If we correct this to two decimal places, using the rule listed in section 2.4, it becomes:

16.32.

This number is now correct to four significant figures.

The same thing can be applied to whole numbers. The number 65 121 is correct to 5 significant figures; the number 65 120 is correct to four significant figures, that is, it is rounded off to the nearest 10. This can be illustrated as follows:

Boxes of oranges are packed to contain 120 oranges. The supervisor is not too concerned if one or two more oranges are added in error. The box is sold as 120 oranges, that is, correct to 2 significant figures.

Again, a bill at a restaurant comes to £51.44. The manager crosses out the 44p and the bill paid is £51.00. The bill is rounded off to 2 significant figures.

The decimal number 0.00716 is correct to 3 significant figures. The two zeros after the decimal point do not count. However, the number:

8.00716

is correct to 6 significant figures.

EXAMPLE 3

Correct the following to 3 significant figures:

(a) 14.76
(b) 8.315
(c) 0.005176
(d) 0.08165
(e) 1778
(f) 3.0061
(g) 92 316

Using the rounding-off rule, we have:

(a) 14.8
(b) 8.32, since the digit immediately preceding is odd
(c) 0.00518
(d) 0.0816 since the digit immediately preceding is even
(e) 1780, i.e. corrected to the nearest 10
(f) 3.01
(g) 92 300, i.e. corrected to the nearest 10

2.6 BASIC OPERATIONS WITH DECIMALS

(a) *ADDITION AND SUBTRACTION*

Addition and subtraction is the same as for whole numbers, with the requirement that the decimal points of the numbers line up; as shown in Examples 4 and 5.

EXAMPLE 4

Add 0.716, 14.4, 2100, 0.0066 ______________

```
   0.716
  14.4
2100.
   0.0066
---------
2115.1226
---------
```

EXAMPLE 5

Subtract 16.71 from 23.4 ______________

For convenience a zero is added to 23.4 to make it 23.40.

```
23.40
16.71
-----
 6.69
-----
```

(b) MULTIPLICATION

The two numbers are multiplied together ignoring first of all the decimal point.

Then,

number of decimal places in the answer	=	sum of decimal places in the two numbers being multiplied

EXAMPLE 6

Multiply 4.21 by 3.7 ______________

Ignoring the decimal points

```
  421
   37
-----
 1263    (× 3)
  2947   (× 7)
-----
15577
-----
```

In 4.21 and 3.7 there are a total of 3 decimal places. Therefore, there are 3 decimal places in the answer, that is,

15.577

(c) *DIVISION*

Division is carried out by first making the dividing number a whole number, at the same time moving the decimal point of the number being divided, by the same number of places.

EXAMPLE 7

Divide 0.2716 by 0.052, giving the answer correct to two decimal places

0.27162 ÷ 0.052

We make 0.052 a whole number by moving the decimal point 3 places in both numbers, i.e. 271.62 ÷ 52.

```
         5.223
     ┌────────
  52 │ 271.620      For convenience a
       260          zero is added after
       ───          the 2
        11 6
        10 4
        ────
         1 22
         1 04
         ────
           180
           156
           ───
            24
           ───
```

The answer is 5.22 correct to two places of decimals.

EXERCISES 2.2

1. Express the following to 2 places of decimals:

(a) 17.664 (b) 5.874

(c) 18.965 (d) 19.935

(e) 0.0765
(f) 3.00712
(g) 4.0016
(h) 8.963
(i) 11.996
(j) 7.775

2. Correct the following to 3 significant figures:

(a) 17.665
(b) 4.317
(c) 5.815
(d) 7.945
(e) 176.31
(f) 184.5
(g) 1853
(h) 19 676
(i) 0.008125
(j) 0.06645
(k) 9.0621
(l) 3.0072

3. Carry out the following operations:

(a) 16.21 + 3.056 + 17.1272
(b) 14.311 + 2.007 − 6.314
(c) 45.76 −21.333
(d) 0.0076 + 0.000441 + 0.721
(e) 21.31 × 14.2
(f) 3.12 × 6.44
(g) 17.44 ÷ 2.6 correct to 2 places of decimals
(h) 43.6 ÷ 6.21 correct to 2 places of decimals

MISCELLANEOUS EXERCISES 2

1. Convert the following fractions to decimals correcting to 3 decimal places where appropriate:

(a) $\frac{3}{8}$
(b) $\frac{3}{11}$
(c) $\frac{2}{5}$
(d) $\frac{4}{7}$
(e) $\frac{3}{13}$
(f) $\frac{6}{7}$
(g) $\frac{7}{8}$

2. Express $\frac{1}{2} + \frac{3}{4} + \frac{5}{9} + \frac{11}{12} - \frac{1}{18}$ as a decimal, correct to 2 places of decimals.

3. Correct the following decimals to the stated requirement:

(a) 0.17614 to 2 decimal figures
(b) 0.01764 to 3 significant figures
(c) 0.471 to 1 decimal figure
(d) 3.0567 to 2 decimal places
(e) 3.0567 to 2 significant figures

4. Carry out the following operations:
(a) 3.167 + 14.21 + 7.3
(b) 16.41 − 9.77
(c) 4.21 × 3.17 correct to 2 places of decimals
(d) 71.62 ÷ 3.51 correct to 2 places of decimals

5. Convert the following decimals to fractions and reduce to the simplest form:
(a) 0.16
(b) 0.24
(c) 0.22
(d) 0.18
(e) 0.92
(f) 0.625

3
PERCENTAGES

3.1 THE MEANING OF PERCENTAGE

Consider the square ABCD in Fig. 3.1 divided into 100 parts. Let 17 of these 100 parts (shown shaded in the figure) be removed. This shaded section is 17 out of the 100 parts. It is said to be 17 per cent, written as 17%.

The percentage is thus seen to be a fraction of 100.

Therefore $17\% = \frac{17}{100}$

Again $20\% = \frac{20}{100}$

We can write percentages as fractions with a denominator of 100, and then simplify the fraction. Thus,

$$60\% = \frac{60}{100} = \frac{3}{5}$$

$$25\% = \frac{25}{100} = \frac{1}{4}$$

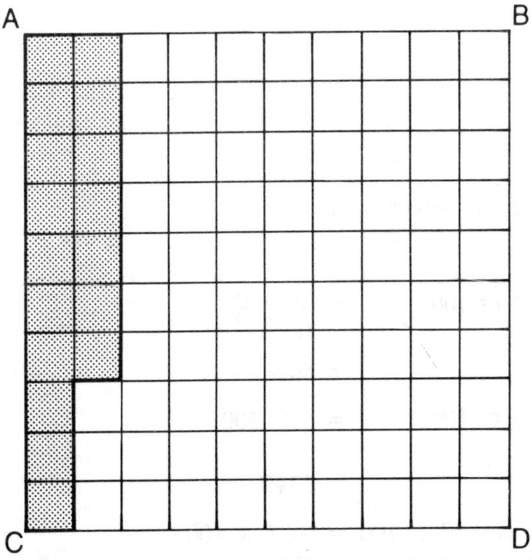

FIG. 3.1 A grid of 100 squares with 17 shaded, i.e. 17%

Again, percentages can be written as decimals:

$$60\% = \frac{60}{100} = 0.6$$

$$33\% = \frac{33}{100} = 0.33$$

$$12.5\% = \frac{12.5}{100} = 0.125$$

$$7\% = \frac{7}{100} = 0.07$$

3.2 TO EXPRESS A FRACTION AS A PERCENTAGE

Let us consider the fraction $\frac{3}{4}$, which means 3 parts out of 4 parts. To write this as a percentage we wish to find how many parts from 100 this is, that is, we want to find

$$\frac{3}{4} \text{ of } 100$$

$$= \frac{3}{4} \times \frac{100}{1}$$

$$= 75$$

Therefore $\frac{3}{4}$ is the same thing as 75 parts from 100 parts, that is

$$\frac{3}{4} \text{ is } 75\%$$

Again $\frac{2}{5}$ is $\frac{2}{\cancel{5}} \times \overset{20}{\cancel{100}}$

$$= 40\%$$

EXAMPLE 1

Express (a) $\frac{4}{5}$ (b) $\frac{1}{3}$ (c) 0.37 as percentages

(a) $\frac{4}{5}$ as a percentage is $\frac{4}{5}$ of 100 $= \frac{4}{\underset{1}{\cancel{5}}} \times \frac{\overset{20}{\cancel{100}}}{1}$

$= 80\%$

(b) $\frac{1}{3}$ as a percentage is $\frac{1}{3}$ of 100 $= \frac{1}{3} \times 100$

$= 33\frac{1}{3}\%$

(c) 0.37 as a percentage is 0.37 of 100 $= 0.37 \times 100$
$= 37\%$

EXERCISES 3.1

1. Express the following as fractions and decimals:

35%	28%	40%
$62\frac{1}{2}\%$	90%	$66\frac{2}{3}\%$
12%	17%	31%

2. Convert the following to percentages:

$\frac{3}{8}$	$\frac{3}{5}$
$\frac{9}{10}$	$\frac{11}{20}$
$\frac{4}{25}$	0.61
0.55	0.816
0.79	0.005

3.3 TO FIND THE PERCENTAGE OF A GIVEN NUMBER

A calculation of this type is as follows. A floor in a hotel has an area of 300 m^2. Of this total area, 40% of the floor is to be carpeted. We want to know the area of carpet required.

We know that 40% is 40 parts from 100 parts, so that an area of 100 m^2 will be covered by 40 m^2 of carpet. To calculate the total area of carpet, we find how many 100s are in 300, that is

$$\frac{300}{100} = 3$$

Each 100 m^2 of floor has 40 m^2 of carpet. Therefore, the total area of carpet is

$$3 \times 40 = 120 \text{ m}^2$$

Therefore, 40% of 300 $= 40 \times \frac{300}{100}$

which is

$$= \frac{40}{100} \times \frac{300}{1}$$

$$= 120 \text{ m}^2$$

The following example illustrates the problem.

EXAMPLE 2

A hotel has 250 bedrooms. It is found that 20% are booked for a whole week. How many bedrooms is this?

Number of rooms booked for the week

$= 20\%$ of 250

$= \frac{20}{100} \times 250$

$= \frac{20}{100} \times \frac{250}{1}$

$= 50$

EXAMPLE 3

In a formal dinner for 300 guests a choice of three main dishes appeared on the menu. The guests chose as follows:

33% chose *Cuissot de Porc Rotï*
21% chose *Poulet Sauté Chasseur*
the remainder chose *Entrecôte Bordelaise*.

Find the number of each dish prepared by the chef.

Cuissot de Porc Rotï: 33% of 300

$= \frac{33}{\cancel{100}_1} \times \frac{\cancel{300}^3}{1} = 99$

Poulet Sauté Chasseur: 21% of 300

$= \frac{21}{\cancel{100}_1} \times \cancel{300}^3 = 63$

Total of the above two: $99 + 63 = 162$

Entrecôte Bordelaise: remainder

$= 300 - 162 = 138$

EXERCISES 3.2

1. Find:

(a) 20% of £400
(b) 35% of 600 m^2
(c) 15% of £45

(d) 22% of 30 kg
(e) 14.5% of 70 m^2

2. The floor area in a hotel is 6000 m^2. 80% of the total area is carpeted. What is the total area of carpet?

3. The floor area of a kitchen is 150 m^2, but only 45% is free of equipment. What is the area of the floor that must be cleaned each day?

4. A cellar has a stock of 45 bottles of German white wine, but only 40% of them are Qualitätswein mit Prädikat. How many bottles of Qualitätswein mit Prädikat are there in stock?
   What is the total value of the stock if the Qualitätswein mit Prädikat wine is £4.50 per bottle and the remainder £2.50 per bottle?

5. The percentage profit on sales of drink is 58%. Calculate the actual profit on sales of £780.

6. 32 kg of lamb was delivered to a kitchen. 6% of the meat was trimmed. How much of the raw meat was waste?

7. A car was bought for £1840 and sold later for 20% less. What was the actual reduction?

8. In the production of fast food meals 0.6% of the meals are discarded. On a day when 4500 meals were produced, how many were wasted?

9. Out of a delivery of 75 kg of meat, 35% was wasted in preparation and cooking. How much meat was left after cooking?
   If the remaining meat was served in 125 g portions, how many portions were served?

10. During one month a hotel used 4500 units of electricity. The following month there was a decrease of 12%. Find the number of units used during the second month.

11. The equipment in a hotel is valued at £15,000. It depreciates at the rate of 10% per annum. Find the value of the equipment at the end of three years.

12. Yogurt contains 6% carbohydrate and 86% water. How much carbohydrate and water is there in 6 kg of yogurt?

13. What is the sum payable for kitchen equipment purchased for £200 less 40% trade discount and a cash discount of 5%?
    (HCIMA intermediate)

3.4 TO FIND ONE QUANTITY AS A PERCENTAGE OF ANOTHER

In this type of calculation we find what percentage one quantity is in respect of another. For example, what percentage is 30 male students out of a class of 60?

First of all, we find the fraction of boys, that is

$\frac{30}{60} = \frac{1}{2}$

Then as in section 3.2

$$\frac{1}{2} \text{ as a percentage } = \frac{1}{2} \text{ of } 100$$
$$= \frac{1}{2} \times 100$$
$$= 50\%$$

This type of calculation is shown in Examples 4 and 5.

EXAMPLE 4

On a particular day a restaurant served 200 meals; 40 meals were served with waiter service, the others were self-service meals. What percentage of the total had waiter service?

First of all the fraction out of the total is found:
Fraction of waiter service meals

$= \frac{40}{200}$

Percentage of waiter service meals

$= \frac{40}{200} \times 100$

$= 20\%$

EXAMPLE 5

The bill for a meal in a restaurant for 4 people was £36, inclusive of wine. The wine bill was £6. What percentage of the total was the wine bill?

Fraction of wine bill to total bill

$= \frac{6}{36}$

$= \frac{1}{6}$

Percentage of the winc bill to total bill

$= \frac{1}{6}$ of 100

$= 16\frac{2}{3}\%$

EXERCISES 3.3

1. Find what percentage:
(a) £6 is of £60
(b) 10 apples are out of 250
(c) 32 electrical units are out of 36 units.

2. 25 kg of beef were delivered to a restaurant. After trimming, 24 kg of saleable meat remained. What percentage of beef was trimmed?

3. The total estimated cost of commodities for an industrial catering company was:

Directors' kitchen	£150
Works kitchen	£540
Pastry section	£140
Still room	£70

Find as a percentage of the total cost, the cost of each section.

4. The sales forecast of sirloin steak in a restaurant was 250 in a week. In practice, the actual sales were 225. What was the percentage reduction of the forecast?

5. Express as a percentage:
(i) out of a box of 144 eggs, 3 are bad,
(ii) in a total of 980 employees, 860 eat lunch in the work's canteen,
(iii) 20 lbs of sausage meat contained 8 lbs of fat.

(Neath C)

6. In comparing the meat costs over two months of a canteen's operation, it was found that:

First month:	Meat costs were 17.5% of the total food costs
Second month:	Meat costs £1100
	Vegetables cost £350
	Dairy costs £250
	Other foods £450

Which of the two months has the greater percentage of meat costs?

7. The following table shows the weight of constituents in 250 g of Cheddar cheese and cottage cheese:

Constituent	Cheddar cheese		Cottage cheese	
	Weight (g)	%	Weight (g)	%
Protein	65		35	
Fat	85		10	
Carbohydrate	0		2.5	
Minerals	7.5		2.5	
Water	92.5		200	

Find the percentage of each constituent in both cheeses.

8. The average temperature of the lounge and dining room of a hotel was set at 65 °F. The outside temperature changed and the temperature of the lounge became 60 °F and the dining room 58 °F. Find the percentage temperature drop in each case.

9. A hotel owner works on average 52 hours per week during the peak season. He finds that his time is roughly divided as follows:

Checking accounts and stock	13 hours
Meeting guests and duty rota in the evenings	$19\frac{1}{2}$ hours
Menu planning with the chef	$6\frac{1}{2}$ hours
Routine inspection of hotel	$3\frac{1}{4}$ hours
Mail and correspondence	$3\frac{1}{4}$ hours
Staff training, recruitment, problems	$6\frac{1}{2}$ hours

What do these duties represent as percentage of the total weekly hours?

10. The costs for a typical week in the hotel office were found to be:

Letters	£24.50
Telephone	£38.25
Paper	£8.25
Photocopying	£29.00

Find what each item is, as a percentage of the total.

3.5 TO FIND THE WHOLE AMOUNT GIVEN A PERCENTAGE

This calculation is illustrated below. It is important to note that the whole amount is 100%.

Of a floor area, 10% is carpeted and this is actually 36 square yards of carpet. What is the total floor area?

10% is actually 36 square yards

1% is $\frac{36}{10}$ square yards (i.e. 10 times less)

100% represents the total area $= \frac{36}{10} \times 100$ (i.e. 100 times more)

$= 360$ sq yds

EXAMPLE 6

The total profit of the grill restaurant of a hotel in one week was £1350. If the percentage profit was 45% of sales, what were the actual sales?

As above, 45% is actually £1350

$1\% = \frac{1350}{45}$ (i.e. 45 times less)

100% represents the total sales $= \frac{1350}{45} \times 100$ (i.e. 100 times more)

$= £3000$

EXAMPLE 7

A firm of mobile caterers quoted for a function at a garden fête. Transport was estimated at £240, which was 15% of the total costs. What was the total cost?

15% was actually £240

$1\% = \frac{240}{15}$

100% represents total costs $= \frac{\cancel{240}^{80}}{\cancel{15}_{\cancel{5}}} \times \cancel{100}^{20}$

$= £1600$

EXERCISES 3.4

1. 15% of a bill came to £24. What was the total bill?
2. 20% of a floor area was 40 m^2. What was the total floor area?
3. A consignment of fish had not been properly refrigerated and 12 boxes representing 6% of the total had to be disposed. How many boxes were there in the consignment?

4. A trainee chef received a bonus of $6\frac{1}{4}\%$ of his annual salary, which amounted to £192. What was his annual salary?

5. A man sold his car for £2000, which was $62\frac{1}{2}\%$ of the sum that he originally paid for it. What was the initial price of the car?

6. A wine cask contained 77 litres after $12\frac{1}{2}\%$ had been removed. Calculate the number of litres in the cask originally.

(Neath C)

7. In one week 1150 meals of fish and chips were served in a canteen. The total cost of fish was £750. If the cost of the fish was 75% of the total, calculate the total cost of these meals.

Hence, calculate the cost per meal to the nearest penny.

(Neath C)

8. The number of salads served in a school canteen was 12% of the total meals served. If the number of salad meals served was 180, find the total number of meals served.

9. During the winter months on average, the number of rooms unoccupied in a hotel was found to be 20%, representing 85 rooms. Find how many bedrooms there were in the hotel.

10. In any given week the hotel manager spent about 25% of his week talking to guests. If this represented 12 hours, what was his total weekly hours?

11. In a Farm Tourism enterprise organised in conjunction with the Tourist Board, the number of weeks in which the accommodation was fully booked was 18, representing 45% of the season. How long was the tourist season?

MISCELLANEOUS EXERCISES 3

1. Write the following percentages as fractions, reducing them to their simplest forms:

(a) 65% (b) 45%
(c) $12\frac{1}{2}\%$ (d) 6%

2. Find the following values:

(a) 20% of 3000 kg
(b) 16% of £450
(c) $27\frac{1}{2}\%$ of 600 m^2
(d) 75% of 520 km

3. In each of the following, find the total amount:
(a) $12\frac{1}{2}\%$ is £400
(b) 30% is 270 ft^2
(c) 18% is 540 miles
(d) 55% is 297 boxes

4. What is:
(a) 66 as a percentage of 88?
(b) £50 as a percentage of £300?
(c) £3.50 as a percentage of £35?
(d) $4\frac{3}{4}$ kg as a percentage of 76 kg?

5. Increase:
(a) 774 by 5%
(b) £390 by 10%
(c) 450 kg by 6%
(d) 45°C by 20%

6. Tiles in a kitchen absorb moisture. In order to test the suitability of a tile, it should not absorb more than 0.5% of its weight after soaking for 12 hours under water. Calculate which of the following would be acceptable, from the three specimens, A, B, C.

	A	B	C
Dry weight	370	382	373
Wet weight	374	384	374

(Oxford)

7. A wholesaler buys an article for £260 and sells it to a retailer and provides the wholesaler with a profit of 30%. Calculate the price that the wholesaler sold the article.

The retailer then sells the article to a customer at a profit of 25%. Calculate the actual cost to the customer.

Find, too, the total profit made on the article based on the cost to the wholesaler.

(GLOSCAT)

8. Dried whole milk powder contains the following constituents by percentage:

Protein	26%
Fat	26%
Carbohydrate	39%
Minerals	5%

Water	3%
Other traces	1%

Find the weight of each constituent in a box of 850 g.

9. Equipment in a works canteen is valued at £15,000. Each year it depreciates 10% of its value. Find the value of the equipment after 3 years.

(GLOSCAT)

10. The chef of a hotel has invested £1000 in a Building Society at a rate of 9%. Find how much will be invested at the end of three years, if the interest each year is retained in the account.

4

PROFIT, STOCK AND VAT

4.1 GROSS AND NET PROFIT: STOCK

(a) *GROSS PROFIT*

Gross profit refers to the profit made on the materials and commodities used. It can be expressed as:

Gross profit = Selling price − Buying (cost) price

(b) *NET PROFIT*

In any business there are other factors which must be taken into account besides the cost of commodities; these are:

labour costs—wages
staff accommodation
staff expenses.

In addition there are other costs called **overheads**:

rates
repairs
heating
depreciation on equipment.

In the profit equation these other costs must be accounted for as well. The profit so made is called the **net profit**, which is:

Net profit = Selling price − Total costs

Note: Gross profit can be calculated for each section of operation, but net profit only makes sense if it is calculated for the whole establishment.

In most commercial enterprises, the percentage profit is calculated from the cost price. In the hotel world, however, percentage profit is always on the selling price.

EXAMPLE 1

A quantity of fish was bought for £25. On the menu the fish course was priced to give a selling price of £40. Find the gross profit and percentage gross profit.

Gross profit = selling price − cost price
= £40 − £25
= £15

Percentage gross profit (as a % of selling price) = $\frac{15}{40} \times 100$
= 37.5%

EXAMPLE 2

The total sales of a restaurant for a week was £3600. The costs were as follows:

Cost price of food	= £1200
Labour costs	= £800
Estimated overheads	= £1000

Find the percentage net profit.

Total costs = £3000
Net profit = sales − total costs
= £3600 − £3000
= £600

Net percentage profit = $\frac{600}{3600} \times 100$

= $\frac{100}{6}$

= $16\frac{2}{3}$%

EXAMPLE 3

A small food firm employs 20 people for 40 hours per week, paying them £3.00 an hour. The weekly overheads are £100.

The firm makes a fixed number of pies per week. The cost of commodities is £500 per week and sales bring in an income of £4000 per week.

Calculate:

(i) gross profit
(ii) net profit
(iii) percentage net profit.

(i) Gross profit = selling price − cost price of materials
= 4000 − 500
= £3500

(ii) Total costs:

Wages of each employee	= 40 × £3	= £120
Total wages bill	= £120 × 20	= £2400
Overheads		= £100
Materials		= £500
Total costs		£3000
Net profit	= £4000 − £3000	
	= £1000	

(iii) Percentage net profit:
Being a food firm and not a hotel,

$$\text{Percentage net profit} = \frac{\text{net profit}}{\text{total cost}} \times 100$$

$$= \frac{1000}{3000} \times 100$$

$$= 33\tfrac{1}{3}\%$$

(c) *STOCK*

In arriving at basic cost of food materials in order to calculate gross percentage profit, one has to take into account cost of purchases and cost of stock used up.

From Fig. 4.1 it is seen that:

Cost of materials used = Initial stock + Purchases − Final stock

For example, in the month of June the following figures were obtained in a restaurant:

(i)

Value of stock at the beginning of the month	Value of stock at the end of the month	Purchase made during the month
£756	£423	£1350

From Fig. 4.1(a)

∴ Cost of materials used = £756 + £1350 − £423 = £1683

Again, consider the following example:

(ii)

Value of stock at the beginning of the month	Value of stock at the end of the month	Purchases
£872	£996	£1260

When the stock value increases, some of the purchases are not used. From Fig. 4.1(b)

$$\therefore \text{Cost of materials} = £872 + £1260 - £996$$
$$= £1136$$

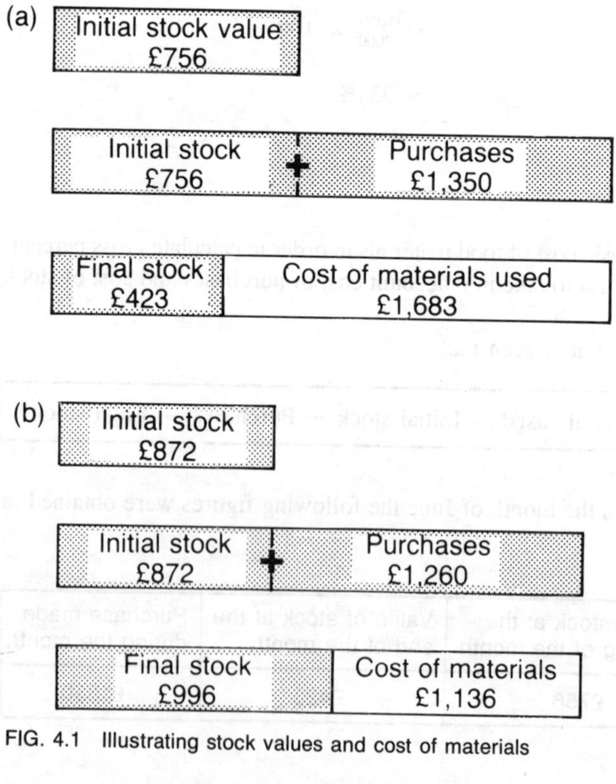

FIG. 4.1 Illustrating stock values and cost of materials

EXAMPLE 4

In the following sales patterns for two months find the gross percentage profit:

	Stock value at beginning of the month	Stock value at end of the month	Purchases	Sales of food and wine
June	£636	£496	£1900	£6450
July	£496	£726	£1800	£5610

(i) For the month of June

Cost of materials used = Initial stock + purchases − final stock

= £636 + £1900 − £496
= £2040

Gross profit = £6450 − £2040
= £4410

Gross percentage profit $= \frac{4410}{6450} \times 100$
= 68%

(ii) For the month of July

Cost of materials used = £496 + £1800 − £726 = £1570

Gross profit = £5610 − £1570 = £4040

Gross percentage profit $= \frac{4040}{5610} \times 100$
= 72%

EXERCISES 4.1

1. In the following table calculate the gross profit as a percentage of the selling price:

	Selling price	Food cost
(a)	30p	12p
(b)	£72	£28.50
(c)	£4700	£1300

(Blackpool and Fylde)

2. If the overheads in Question 1 are respectively:

(a) 5p
(b) £11.50
(c) £1200,
calculate the net percentage profit in each case.

3. A quantity of ground coffee was purchased for 90p per $\frac{1}{4}$ kg. $\frac{1}{4}$ kg makes approximately 30 cups of black coffee. If the price on the menu for coffee is 20p per cup, what was the gross profit per $\frac{1}{4}$ kg? Calculate also the percentage gross profit.

4. Compare the gross profit and gross percentage profit of the following sales:
(a) 10 meals costing £2.40 each and selling for £6 each
(b) 10 meals costing £4.00 each and selling for £10 each
(c) 100 meals costing 60p each and selling for £1.20 each.

(Salisbury)

5. If you sell food at four times its cost, calculate your gross percentage profit.

6. For the kitchen a certain item was bought for £1.20 and sold for £3.00. Calculate the gross percentage profit.

7. In the following table d'hôte menu, the costs are:

Food cost	£4.50
Labour costs	£4.00
Overheads	£3.50
Meal price	£15.00

Calculate:
(i) the gross percentage profit
(ii) the net percentage profit.

8. A small food firm employs 10 people for 38 hours per week, paying them £3.00 an hour. In addition, each worked an average of 5 hours overtime per week at a 'time and a half' rate per hour. The weekly overheads were £1450.

If the cost of materials was £350 per week and the sales brought an income of £3500, calculate:
(i) gross profit and gross percentage profit
(ii) net profit and net percentage profit.
(*Note*: This is a food manufacturer and not a hotel.)

9. In a typical week, a restaurant had the following costs:

Food costs	£4000
Cleaning and administration materials	£500

Labour costs	£6000
Overheads	£2500
Income (food)	£17,000

Calculate the net percentage profit.

10. The profit from food sales in a hotel was £9 per meal, the menu price being £12. On average 180 meals were served each day. If the labour costs were £3000 for the week and overheads £2000, calculate:

(i) Gross percentage profit

(ii) Net percentage profit.

11. In the following table, find the gross percentage profit for each month (correct to 2 significant figures).

	Opening Stock	Closing Stock	Purchases	Sales
August	£490	£790	£1200	£6540
September		£430	£900	£4770
October		£680	£1500	£3880

4.2 VALUE ADDED TAX

(a) On most goods that we buy, a tax of 15% is added, called value added tax—VAT. To calculate any VAT, the calculation is the same as in section 3.3, that is, the calculation of 15% of a given basic price.

EXAMPLE 5

The floor of the restaurant is carpeted. The cost is £2200. VAT is added at 15%. Calculate:

(i) the amount of VAT

(ii) the total cost to the restaurant owner.

(i) VAT = 15% of £2200

$= \frac{15}{100} \times 2200$

= £330

(ii) Total cost = Basic cost + VAT

= £2200 + £330

= £2530

(b) Now we come across an awkward little calculation. If a customer buys a lunch at a restaurant, the price includes the VAT. If he wishes to find how much VAT is included a calculation as follows is necessary. It is necessary to work backwards from the actual bill.

The total cost of the lunch for 4 people was £62, inclusive of VAT at 15%. The difficulty is, of course, that the 15% is on the basic price, not on the £62.

i.e. Basic price + VAT = £62

> The trick is to let the basic price be 100%.

∴ Basic price + VAT = Selling price
100% + 15% = 115%

Thus, 115% makes £62

1% makes $£\frac{62}{115}$

15% VAT makes $£\frac{62}{115} \times 15$

= £8.09

The VAT in the bill was £8.09.

EXAMPLE 6

An electric service counter was bought for £5600, which included VAT at 15%. The VAT could be claimed back by the purchaser. What is the VAT?

Basic cost + VAT = £5600

Let the basic cost be 100%

100% + 15% = £5600

i.e. 115% makes £5600

1% makes $£\frac{5600}{115}$

15% makes $£\frac{5600}{115} \times 15$

= £730.43

∴ VAT is £730.43

EXERCISES 4.2

1. The basic cost of linen from a wholesaler was £340. The cost to the hotelier had VAT added at 15%. What was the cost to the hotelier?

2. An article cost £45 + VAT at 15%. What was the purchase cost of the article?

3. A dishwasher was bought for a basic cost of £750 + VAT at 15%. What was the price actually paid for it?

4. Inclusive of VAT at 15% the bill for an article was £2.99. How much was VAT?

5. A meal in a restaurant for four people cost £45, including VAT. How much VAT was paid?

6. Complete the following table by filling in the blank spaces:

Basic selling price of an article	£3.50	£120.00	£630.00	—	—
Selling price + VAT at 15%	—	—	—	£18.50	£172.50

7. The bill in a restaurant for a group of people reads as follows:

Soup	15 @	20p
Fish	12 @	95p
Duck	4 @	£3.50
Steak	8 @	£3.75
Sherry trifle	12 @	45p

Calculate:
(a) the total cost of the meal
(b) the amount of VAT included in the bill if VAT is 15%.

8. What was the original amount of a hotel bill for £70.84 which includes 10% service charge and 15% VAT?

4.3 TO CALCULATE THE SELLING PRICE

If the basic cost of an article is known and the percentage profit is stated, then the selling price can be calculated. However, it is important to note two distinct cases:

(i) *All industries other than the catering industry where selling price = basic price + % profit on the basic price.* The calculation is quite simple and is illustrated as follows:

The cost price (basic price) of a flambé lamp and trolley is £340. The supplier makes 25% profit. At what price does he sell it to a customer?

Profit on the lamp = 25% of £340
= £85

∴ Selling price = £340 + £85
= £425

(ii) *The catering industry.* The difficulty now is that in the catering industry, the percentage profit is based on the selling price and not on the basic price.

i.e. selling price − % profit on selling price
= cost price

To calculate the selling price we use the same trick again as for the VAT problem in section 4.2(b), that is, let the selling price = 100%.

For example, if the basic costs of a restaurant in a given month are £8500 and profit is to be 60%, what is the selling price for that month, that is, the monthly income?

Now let the selling price be 100%.

∴ Selling price − profit on selling price = basic costs

100% − 60% = £8500

∴ 40% = £8500

1% = $\frac{8500}{40}$

100% (selling price) = £$\frac{8500}{40} \times 100$

= £21 250

∴ The selling price = £21 250

EXAMPLE 7

The gross percentage profit on a meal is planned at 40%. If the basic cost of a set menu is £6 per cover, what must it be priced at to make 40% profit (on the selling price of course)?

Let the selling price be 100%.

Thus, 100% − gross percentage profit = cost price
100% − 40% = cost price
60% = cost price of £6

1% = $\frac{6}{60}$

$$100\% \qquad = \tfrac{6}{60} \times 100$$

$$= £10$$

Therefore the meal costs £10 per cover.

EXERCISES 4.3

1. Calculate the selling price of the following:
(a) Electric fire: cost price = £24; profit = $33\frac{1}{3}\%$
(b) Car: cost price = £5000; profit = 25%
(c) Restaurant furniture: cost price = £1500; profit = 22%.

2. Calculate the selling price of the following items in the catering industry (profit is on the selling price):
(a) Bed and breakfast: cost price = £21; profit = 40%
(b) Dinner in the restaurant: cost price = £9; profit = 55%.

3. The total running cost of a hotel per week is £3000. If the directors expect 40% profit, calculate the income that the manager must achieve per week.

4. The bill for food and commodities in a cafe per week is £450. The proprietor tries to achieve a gross profit of 60%. Find the total income per week from selling the food.

EXAMPLE 8

Purchases and issues of potatoes were recorded as follows:

Purchases		Issues				
Opening stock	80 lb @ 4p/lb	Mon	Tues	Wed	Thurs	Fri
Week 1	56 lb @ 5p/lb	20	25	40	20	30
Week 2	112 lb @ $4\frac{1}{2}$p/lb	25	30	45	30	20
	56 lb @ $4\frac{1}{2}$p/lb					
Week 3	56 lb @ 4p/lb	20	25	25	20	20
	56 lb @ $4\frac{1}{2}$p/lb					
Week 4	112 lb @ 5p/lb	25	35	40	30	40
	56 lb @ $5\frac{1}{2}$p/lb					

Deliveries are made on Monday morning each week; if a second delivery is required, it arrives on Thursday morning.

(a) Determine the actual total price of the issues.

(b) Calculate the amount and value of closing stock.

(HCIMA intermediate)

It is essential to lay out the problem clearly and logically.

Notes

1. The total issues are the sum of the figures listed under the day of the week for any given week.
2. The issues use up the number of pounds in store first of all, at the cost price of these potatoes. Then the issues use up the number of pounds delivered on the Monday, at these quoted prices; then those delivered on Thursday.
3. The closing stock is given by

$$\text{Closing stock} = \text{Opening stock} + \text{purchases} - \text{issues}$$

Week	Opening stock	Purchases	Total issues	Issue prices	Issue value	Closing stock
1	80 lb @ 4p	56 lb @ 5p	135 lb	80 lb @ 4p 55 lb @ 5p	£3.20 £2.75	1 lb @ 5p
2	1 lb @ 5p	168 lb @ $4\frac{1}{2}$p	150 lb	1 lb @ 5p 149 lb @ $4\frac{1}{2}$p	£0.05 £6.70$\frac{1}{2}$	19 lb @ $4\frac{1}{2}$p
3	19 lb @ $4\frac{1}{2}$p	56 lb @ 4p 56 lb @ $4\frac{1}{2}$p	110 lb	19 lb @ $4\frac{1}{2}$p 56 lb @ 4p 35 lb @ $4\frac{1}{2}$p	£0.85$\frac{1}{2}$ £2.24 £1.57$\frac{1}{2}$	21 lb @ $4\frac{1}{2}$p
4	21 lb @ $4\frac{1}{2}$p	112 lb @ 5p 56 lb @ $5\frac{1}{2}$p	170 lb	21 lb @ $4\frac{1}{2}$p 112 lb @ 5p 37 lb @ $5\frac{1}{2}$p	£0.94$\frac{1}{2}$ £5.60 £2.03$\frac{1}{2}$	19 lb @ $5\frac{1}{2}$p
			Total price of issues		£25.95$\frac{1}{2}$	

Value of closing stock = £1.04$\frac{1}{2}$

MISCELLANEOUS EXERCISES 4

1. A firm supplied the following goods:

18 cups @ £1.27
9 saucers @ £0.82
14 tea plates @ £1.05
15 dinner plates @ £0.97

The firm gives trade discount of 20% and a further discount of $2\frac{1}{2}$% for prompt payment. VAT is then added at 15%. Calculate the final bill, assuming prompt payment.

(GLOSCAT)

2. A small food firm has weekly overheads of £288. It employs 16 people for 40 hours per week, paying them £2.10 an hour.

The firm makes 1800 dozen pies per week. Ingredients cost 26p per dozen. If the pies sell at £1.26 per dozen, find the net percentage profit per week.

(*Note*: The percentage profit in a food firm is

$$\frac{\text{profit}}{\text{basic costs}} \times 100)$$

(Oxford)

3. Calculate the VAT on the cost of an oven, whose basic price is £1400. VAT is charged at 15%.

4. The bill for a luncheon was £64 for four people. The VAT could be reclaimed. How much was the VAT if it was charged at 15%?

5. The weekly cost of a Motel, including overheads was found to be on average £4500. The net profit was planned to be 55%. Calculate the weekly income required.

6. The following statement summarises the trading results of C. Charles, a restaurant owner, for 1977:

	£	£
Sales		40,000
less cost of sales		16,000
Gross profit		24,000
less labour costs	10,000	
overheads	8,000	18,000
Net profit		6,000

During 1978 Charles hopes to expand the business and estimates that sales will increase to £60,000 with gross profit being maintained at the same rate.

To handle the increased sales, he expects to employ two more staff, paying each of them £1500. Of the overheads, one half of the existing figure would not increase but the other half would increase at the same rate as sales.

Prepare a statement of estimated profit or loss for 1978 in the same form as that for 1977.

(Hint: gross profit is increased at the same percentage as the increased sales)

(HCIMA intermediate)

7. The following figures were extracted from the food records of a restaurant:

	Opening stock	Closing stock	Purchases	Sales
March	£570	£820	£1270	£2200
April	—	£740	£ 984	£2000
May	—	£590	£1036	£2100

(a) What is the difference in the gross profit percentages for the different months?
(b) What would have been the effect of an increase of 7% in raw material costs for April on the gross profit percentage?
(c) What is the gross profit for the quarter?

(HCIMA intermediate)

8. Where opening stock is £500, closing stock is £700, sales £14,500 with a 60% gross profit on sales, what is the value of purchases?

(HCIMA intermediate)

9. Purchases and issues of sausages were recorded as follows for one month:

	Purchases	Issues
Opening stock—29 lb, cost 43p per pound		
Week 1	40 lb at 45p/lb	60 lb
Week 2	48 lb at 43p/lb	40 lb
Week 3	25 lb at 46p/lb	
	30 lb at 42p/lb	60 lb
Week 4	40 lb at 40p/lb	50 lb

Issues are priced on a First-in First-out basis.
What is the amount and value of the closing stock?

(HCIMA intermediate)

10. The information below was extracted from the records of a residential conference centre for May 1978:

	£
Stock at May 1st	365
Stock at May 31st	283
Sales	1620
Wages	490
Perishable purchases	411
Non-perishable purchases	437
Staff meals	20
Cleaning materials	18
Heat, light, power	83
Linen hire	19
Stationery	7
Petty cash	31
Share of rent and rates	45

(a) Classify the items in the elements of cost and express each element as a percentage.

(b) Calculate the gross profit or loss percentage.

5

THE ELECTRONIC CALCULATOR AND THE COMPUTER

5.1 THE DISPLAY AND KEYS OF AN ELECTRONIC CALCULATOR

Many aids have existed to help us to carry out calculations. In primitive times counting was carried out with stones and at some time long ago the abacus appeared, which counted items in tens using beads on horizontal strings.

Multiplication tables are aids to carrying out calculations, and so are logarithm tables.

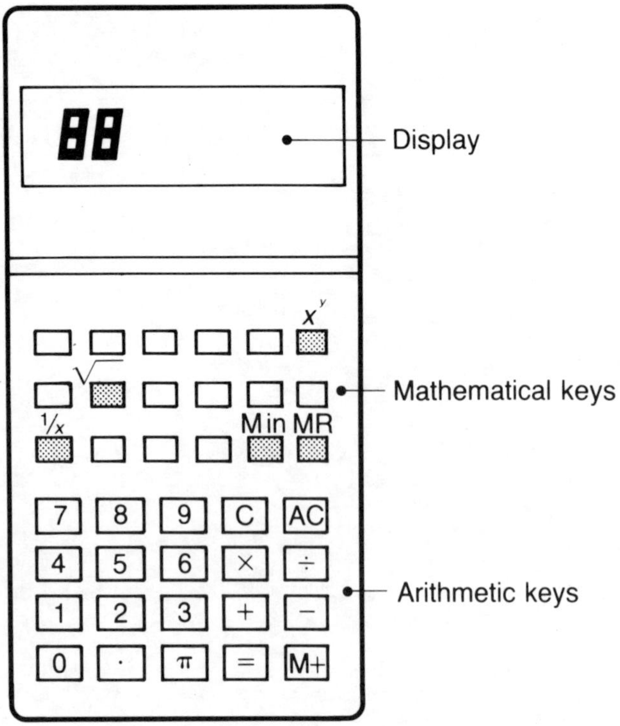

FIG. 5.1 The electronic calculator

Nowadays the electronic calculator is a very powerful aid to carrying out all sorts of complicated calculations. A typical electronic calculator is shown in Fig. 5.1.

It consists of a display, where the numbers appear, a set of arithmetic keys [0] to [9], a set of mathematic keys, including reciprocal [$\frac{1}{x}$] and [x^y] as shown. In addition, there are operation keys, [+], [−], [×], [÷], [=] and the memory keys [M].

The display is really a series of light emitting diodes, (LEDs), which are miniature fluorescent tubes made up into figures of eight as shown in Fig. 5.2(a), each figure of eight being composed of seven LEDs. Only part of each figure of eight will light up to represent any of the figures 0 to 9, apart from 8, of course, where all seven light up. The figures on diplay then have a rectangular outline, as shown in Fig. 5.2(b).

It is important to note that [C] and [AC] are keys for cancelling numbers which need to be erased. [C] will only cancel the last number entered into the calculator. [AC] will clear everything.

Figure 5.2(a) shows the diodes that light up to show 3; 5.2(b) shows the display of the numbers 0 to 9.

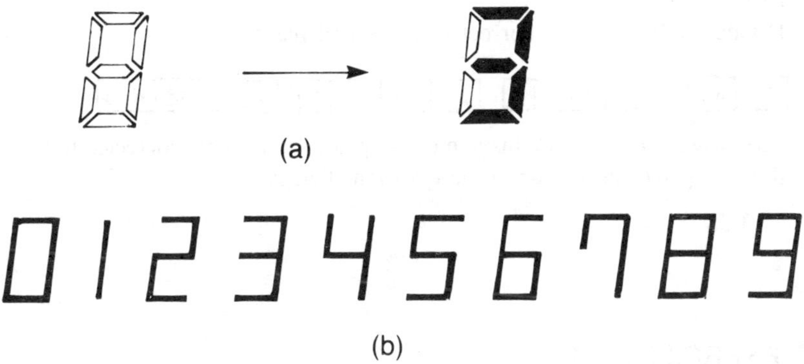

FIG. 5.2 The display figures

5.2 ARITHMETIC OPERATIONS

(a) *ADDITION AND SUBTRACTION*

Two numbers are added using the [+] key. The keys are pressed in the order shown below to add 6 and 5.

[6] [+] [5] [=] [11]

This type of calculation may be extended to larger numbers and those involving decimals. In order to add 63.54 and 23.16, we have:

[6] [3] [·] [5] [4] [+] [2] [3] [·] [1] [6] [=] 86.7

Subtraction follows the same process, using now however, the [−] key, e.g. 9 − 7 is:

[9] [−] [7] [=] 2

(b) *MULTIPLICATION AND DIVISION*

Using the same procedures as above, using the [×] and [÷] keys, we have the following examples.

EXAMPLE 1

Multiply 13.61 by 2.41.

[1] [3] [·] [6] [1] [×] [2] [·] [4] [1] [=] 32.8001

EXAMPLE 2

Divide 76.49 by 22.6, correct to 2 decimal places.

[7] [6] [·] [4] [9] [÷] [2] [2] [·] [6] [=] 3.3845133

The answer is not left to this number of places and is now corrected to two decimal places asked for in the question, that is:

3.38

EXERCISES 5.1

Using the calculator, evaluate the following. Correct the answers to two decimal places, where applicable.

1. 3.6 + 2.7
2. 4.166 + 6.007
3. 18.47 − 2.49
4. 32.091 − 28.66
5. 16.31 × 4.66
6. 17.44 × 0.036
7. 27.2 ÷ 6.64
8. 0.063 ÷ 0.0047

(c) *MIXED OPERATIONS*

The procedures described so far can easily be applied to longer calculations.

Consider the following calculations:

EXAMPLE 3

Evaluate 6.72 + 3.62 + 4.41

We enter:

[6] [·] [7] [2]

[+] [3] [·] [6] [2]

[+] [4] [·] [4] [1]

[=] 14.75

Note that repeated addition accumulates in the working memory until the [=] key is pressed. Once this key is pressed the calculation recommences with the next number 'keyed in'.

EXAMPLE 4

Evaluate 3.16 × 2.41 × 7.66

We enter the numbers as follows:

[3] [·] [1] [6]

[×] [2] [·] [4] [1]

[×] [7] [·] [6] [6]

[=] 58.335496

The mixed calculations can be carried out quite easily, as long as the rules of precedence are obeyed. These were stated in section 1.1.

In calculation 3 + 5 × 4, the rules of precedence determine that the multiplication is carried out first, that is, 5 × 4 = 20, then 20 + 3 = 23. On the calculator we have:

[5] [×] [4] [+] [3] [=] 23

If we add 3 + 5 first on the calculator, that is:

[3] [+] [5] [×] [4] [=] 32

we are really calculating (3 + 5) × 4.

EXERCISES 5.2

Using the calculator and remembering the rules of precedence evaluate the following correct to two decimal places, where applicable:

1. $16.2 + 3.41 \times 2.67$
2. $3.71 - 1.88 \div 0.773$
3. $(4.66 + 3.89) \times 2.783$
4. $6.71 + 2.64 \times 3.12 - 4.17$
5. $3.26 - 1.44 \times 0.62 + 8.47$

5.3 OPERATIONS INVOLVING THE INTERNAL MEMORY

Numbers can be added together into the working memory as shown in section 5.2(c). However, if the results of a series of calculations are required to be added it is necessary to use the internal memory. Any number on the display can be added to the internal memory by pressing [M+] key. This total in the internal memory **cannot** be erased using the [C] or [AC] keys. To find the total in the internal memory [MR] must be pressed, which brings the total into the display.

To erase the internal memory, put [0] into the display and press the [Min] key. In some calculators the internal memory is erased using the [MC] key.

The use of the internal memory is shown in the following examples.

EXAMPLE 5

Evaluate $3 \times 6 + 4 \times 8 + 7 \times 2 + 6 \times 3$

Each of the multiplications are evaluated and added into the internal memory.

[3] [×] [6] [=] 18 [M+]

[4] [×] [8] [=] 32 [M+]

[7] [×] [2] [=] 14 [M+]

[6] [×] [3] [=] 18 [M+]

82 [MR]

EXAMPLE 6

$$\text{Evaluate } \frac{6.31 \times 4 + 3}{86.5 \div 29 + 4.1}$$

In this type of calculation it is necessary to work out the denominator first and place it in memory.

$$
\begin{aligned}
86.5 \div 29 + 4.1 &= 86.5 \div 29 \\
&= 2.9828 + 4.1 \\
&= 7.0828
\end{aligned}
$$

[M+]

$$6.31 \times 4 + 3 = 28.24$$

We now require $28.24 \div 7.0828$

[28.24] [÷] [MR] [7.0828] [=] [3.987]

EXERCISES 5.3

Evaluate the following making use of the memory of the calculator, correct to three places of decimals, where applicable:

1. $7.12 \times 3.67 + 8.41 \times 0.617 + 0.843 \times 3.611$

2. $\dfrac{7.12 \times 8.7 \times 4.1}{26.2} - 3.4 \times 2.7 + 20.631 \div 5.44$

3. $(7.76 - 6.32) \times 7.72 + 88.63 \div 24.4$

4. $\dfrac{7.6 \times 3.4 + 8.2 \times 1.6}{2.2 \times 1.6 - 0.6 \times 0.39}$

5.4 RECIPROCAL AND POWER KEYS, SQUARE ROOT KEY

(a) The reciprocal of a number such as 5 is $\frac{1}{5} = 0.2$. Using the reciprocal key [$\frac{1}{x}$] on the calculator, the reciprocals of numbers can be obtained very quickly.

For example, the reciprocal of 2.25 is found as follows:

[2] [·] [2] [5] [$\frac{1}{x}$] [0.4444]

(b) Using the power key [x^y] the power of any number may be found.

We know that $5^2 = 25$; but $(5.21)^2$ is more difficult. With a calculator, however:

[5] [·] [2] [1] [x^y] [2] [=] [27.1441]

(c) Square roots are always required and all that is necessary to do is key in the required number and press the [√]; for example, √27.27 is obtained as:

[2] [7] [·] [2] [7] [√] [5.2220686]

5.5 CONSTANT MULTIPLICATION

Sometimes it is necessary to multiply a large set of numbers by the same number. It can save time to use the constant multiplication procedure. For example, suppose we wish to multiply a series of numbers by 4.5.

First of all, we set up the calculator to do this. Simply, the [×] key is pressed twice, when the letter K appears in the display, i.e. for our multiplication by 4.5, we have:

[4] [·] [5] [×] [×]

To multiply any number, say 16, by 4.5 all we now have to do is key in 16 and press the [=] key:

[1] [6] [=] [72]

Any other number may be multiplied by 4.5 in the same way. Pressing [×] again removes the constant multiplication instruction.

This facility is particularly useful in calculating percentages such as VAT for example, which has a constant percentage. This is shown in Example 7.

EXAMPLE 7

Calculate VAT at 15% on each of the items in the following list:

Item	Cost price	VAT at 15%
1 Leather goods	£54.50	
2 Table	£199.00	
3 Glassware	£64.80	

VAT is 15% of cost price, that is $\frac{15}{100} = 0.15$. This 0.15 is 'keyed in' as a constant multiplying number.

0.15 [×] [×]

The VAT is now found:

[5] [4] [·] [5] [0] [=] [8.17]

[1] [9] [9] [.] [0] [0] [=] [29.85]

[6] [4] [.] [8] [0] [=] [9.72]

5.6 APPROXIMATIONS

Using a calculator, as with any other operation, errors are likely to occur. Pressing the wrong key is easily done. The question that then arises, when a calculation has been completed, of how confident it is possible to be about the correctness of the answer.

The method to adopt is that of making approximations. Consider the following calculations:

$$\frac{16.72 \times 192.6}{47.2 \times 8.89}$$

If we adjust each number to multiples of tens, or hundreds, this reduces to:

$$\frac{20 \times 200}{50 \times 10} = 8$$

Compare this now with the calculation completed correctly with a calculator, which is 7.67. The two results are quite close together.

Therefore, if an approximation gives a result which is similar to the calculator result, one can be reasonably confident that the latter result is correct.

EXAMPLE 8

Make an approximate calculation of:

$$\frac{17.62 \times 131.56 \times 0.87}{62.7 \times 47.6}$$

and compare it with the result worked out on a calculator.

Approximate result is:

$$\frac{20 \times 150 \times 1}{60 \times 50} \simeq 1$$

Result using a calculator $= 0.68 \simeq 1$

The two results are compatible, so that it is possible to be confident that 0.68 is a correct result.

EXERCISES 5.4

Evaluate the following using approximations and compare the results with calculations carried out using a calculator:

1. 36.77×0.761

2. $\dfrac{196.5}{48.6}$

3. $\dfrac{17.77 \times 8.66 \times 32.9}{5.9 \times 64.7 \times 7.72}$

4. $\dfrac{0.034 \times 167.6 \times 6.77}{19.67 \times 0.812 \times 26.77}$

5. $\dfrac{21.4 \times 0.0086 \times 736.6}{18.2 \times 0.027 \times 89.6}$

5.7 THE COMPUTER

The computer will allow calculations to be carried out much more efficiently than the electronic calculator. However, the most important function of the computer is holding large quantities of data, such as names of people, ages, addresses of firms and companies, and so on. Not only does it store large quantities of information, but it can extract information from this store in

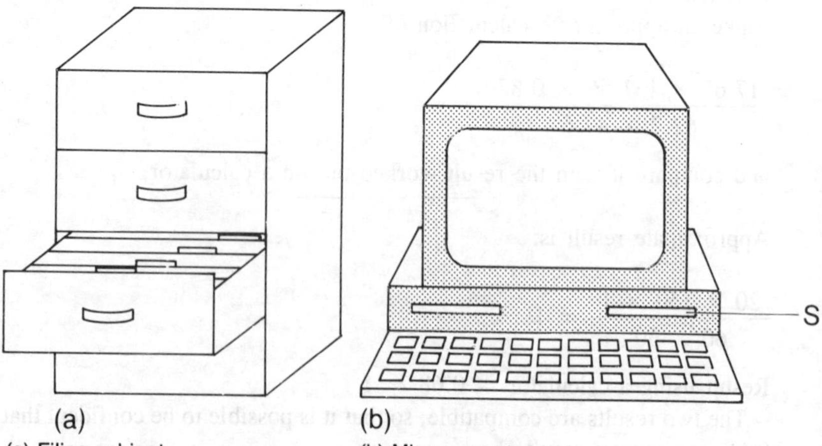

(a) Filing cabinet (b) Microcomputer

FIG. 5.3 Storage of information (data)

this store in a variety of forms and sort the information into alphabetical order or numerical order.

Nowadays, computer programming, that is, the preparation of a set of instructions for the computer, is a specialist exercise. For most of us, the computer with its program is a tool which we use to make our work more efficient.

Each computer has its own operating system, which is the set of instructions built into it by the manufacturer. Into the computer will then be fed a computer program, which is a further set of instructions to suit the particular type of work of the operator. Examples of these programs are word processing, database, spreadsheet, accounting.

(i) The word processing program allows letters and reports to be typed into the memory rather than on a sheet of paper. This work can then be printed out on to paper when given the appropriate instruction. The advantage of the word processor is that sentences and paragraphs can be added, deleted and modified so easily. This is especially important for long reports which require several modifications.

(ii) Database is a program for the storage of information which will be dealt with in more detail below.

(iii) Spreadsheet is a program for performing calculations and is especially useful in forward planning.

(iv) Accounting is a program for keeping income and expenditure in the form of a balance sheet.

5.8 DATABASE

We can think of a database program as an electronic filing cabinet as shown in Fig. 5.3. Information is stored on a floppy disc or a hard disc.

Let us consider how it can be used for storing records of personnel working in a large hotel chain. Let it be assumed that the information shown in Fig. 5.4 is required at the headquarters of the company.

Using a manual system, this card could be completed for each employee and stored in the filing cabinet. The storage could be in alphabetical order, or grouped in employees per hotel, or again by type or status of employment, that is, all the managers together, all the porters together, and so on.

In this way, the information about all the workforce is available. In order to extract information, say everyone over the age of 50, each card has to be examined, which takes a lot of time to go through all the cards. Again, imagine that the cards are stored in strict alphabetical order, and suddenly, a list of employees groups by hotels is required. This again requires a sort of all the cards.

EMPLOYMENT CARD

Surname ______________ Christian names ______________

Home address ______________________________

____

____

Tel. no. ________________ Date of birth ______________

- -

Hotel where employed

Type of employment

Grade

Salary

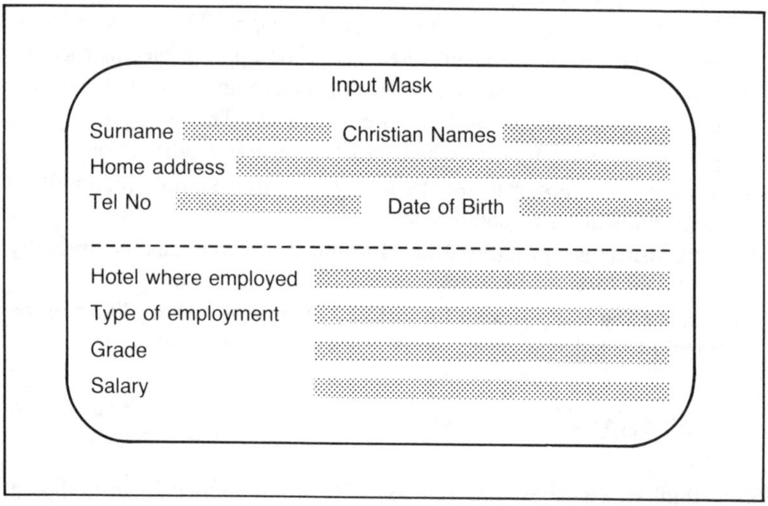

FIG. 5.4 The input mask for staff records, which appears on the screen

Using a computerised system, all this type of sorting and extraction can be done automatically.

The database program will allow the creation of a masterfile corresponding to the information listed on the employment card. This information is then stored on disc via an input mask on the screen of the computer, as shown in Fig. 5.4. It is typed in via the keyboard in the shaded boxes on the screen. The records can be put into computer in any order and stored on discs which are inserted in the slots S (Fig. 5.3b).

Once the records are all stored on the disc, the database, given the correct command, will sort out the records in any order required by the company. The data can then be printed out in the order required.

6

CONVERSION CHARTS AND GRAPHS

6.1 MEASUREMENTS IN THE BRITISH AND METRIC SYSTEMS

Measurements are made and are needed in recording:

length
area
volume (capacity)
mass (often called weight)
energy
temperature

So that sizes can be recorded in each of these, we must first of all agree on a basic size in order to make a measurement. This basic size is called the **unit of measurement**. For example, length can be measured in feet (abbreviation ft). 1 ft is the basic unit.

Two systems of units have been, and are still in common use — the British system and the metric system. The metric system was largely used on the Continent of Europe, with the British system in Britain and the Commonwealth. As the nations became more interdependent, it was decided to use a common system which was largely the metric system. It was called the Système Internationale (SI).

Nevertheless, both the metric system and the British system are still used in the catering industry and consequently it is important to be able to convert from one system to another with ease.

In the British system, changing from one unit to another involves many different conversion factors, such as:

12 in = 1 ft,   3 ft = 1 yd

In the metric system the conversion factor is, generally 1000, e.g.

1000 mm = 1 m.

These factors of conversions are shown in Appendix 1. The following however gives the conversions from metric to British and vice versa.

(a) *MEASUREMENT OF LENGTH*

The table gives the various units in common use in both systems. The accepted abbreviations are shown in brackets.

British units		Metric units	
inch	(in)	millimetre	(mm)
foot	(ft)	centimetre	(cm)
yard	(yd)	metre	(m)
mile	(m)	kilometre	(km)

The conversion from one system to another is given below:

(i)

0.0394 in	←—• mm	1	in •—→	25.4 mm

This is interpreted as: 1 in = 25.4 mm
1 mm = 0.0394 in

(ii)

3.28 ft	←—• m	1	ft •—→	0.305 m

(iii)

1.09 yd	←—• m	1	yd •—→	0.914 m

(iv)

0.621 mile	←—• km	1	mile •—→	1.61 km

(b) *MEASUREMENT OF AREA*

British units		Metric units	
square inch	(in^2)	square millimetre	(mm^2)
square foot	(ft^2)	square metre	(m^2)

$10.8\ ft^2$	←—• m^2	1	ft^2 •—→	$0.0929\ m^2$

(c) *MEASUREMENT OF VOLUME (CAPACITY)*

British units		Metric units	
fluid ounces	(fl oz)	millilitre	(ml)
gill		centilitre	(cl)
pint	(pt)	litre	(l)
gallon	(gal)	metre^3	(m^3)

(i)

35.2 fl oz	←• ℓ	1	fl oz •→	28.4 ml

(ii)

1.76 pt	←• ℓ	1	pt •→	0.568 l

(iii)

0.22 gal	←• ℓ	1	gal •→	4.55 l

(iv)

0.007 gill	←• ml	1	gill •→	1.45 ml

(d) *MASS (WEIGHT)*

British units		Metric units	
ounce	(oz)	gram	(g)
pound	(lb)	kilogram	(kg)

(i)

0.0353 oz	←• g	1	oz •→	28.4 g

(ii)

2.21 lb	←• 1 kg	1	lb •→	0.454 kg

(e) *ENERGY*

British units	Metric units
calorie (cal)	Joule (J)

0.239 cal	← J	1	cal →	4.18 J

(f) *TEMPERATURE*

British units	Metric units
°Fahrenheit (°F)	°Celsius (°C)

There is no single conversion factor, but the following method is easy to remember:

(i)

°F → °C		+40		$\times\frac{5}{9}$		−40

The rule is add 40, multiply by $\frac{5}{9}$, subtract 40, for example:

$$\begin{aligned} 68°\text{F: add } 40 &= 108 \\ \text{multiply by } \tfrac{5}{9} &= 60 \\ \text{subtract } 40 &= 20 \end{aligned}$$

i.e. 69°F is equal to 20°C.

(ii)

°C → °F		+40		$\times\frac{9}{5}$		−40

The rule is: add 40, multiply by $\frac{9}{5}$, subtract 40. For example:

$$\begin{aligned} 40°\text{C: add } 40 &= 80 \\ \text{multiply } \tfrac{9}{5} &= 80 \times \tfrac{9}{5} \\ &= 144 \\ \text{subtract } 40 &= 104°\text{F} \end{aligned}$$

6.2 CONVERSION TABLES

(a) *CONVERSION OF INCHES TO MILLIMETRES*

Whenever a conversion is often needed it is useful to set up a conversion table, so that reference can be made to it readily. We shall produce a table for the conversion of inches to millimetres. From section 6.1 we know that the conversion factor is:

$$1 \text{ in} = 25.4 \text{ mm}$$

Therefore: $2 \text{ in} = 2 \times 25.4 \text{ mm}$

$$4 \text{ in} = 4 \times 25.4 \text{ mm}$$

We can carry out the conversion using the 'constant multiplication' operation on the calculator, as shown in Chapter 5; 25.4 is the constant multiplier; therefore we have:

[2] [5] [·] [4] [×] [×]

Hence converting 2.2 in to mm is obtained as:

[2] [·] [2] [=] 55.88

Conversion chart, inches to millimetres, in steps of 0.1 in										
inches	0.0	0.1	0.2	0.3	0.4	0.5	0.6	0.7	0.8	0.9
0	0.0	2.5	5.1	7.6	10.2	12.7	15.2	17.8	20.3	22.9
1	25.4	27.9	30.5	33.0	35.6	38.1	40.6	43.2	45.7	48.3
2	50.8	53.3	55.9	58.4	61.0	63.5	66.0	68.6	71.1	73.7
3	76.2	78.7	81.3	83.8	86.4	88.9	91.4	94.0	96.5	99.1
4	101.6	104.1	106.7	109.2	111.8	114.3	116.8	119.4	121.9	124.5
5	127.0	129.5	132.1	134.6	137.2	139.7	142.2	144.8	147.3	149.9

TABLE 1

The conversion table of inches into millimetres is shown in Table 1. The box at any intersection of row and column, as shown by the shaded parts in the chart, represents the inches (on the left-hand column) and decimal of an inch (shown by the row on top of the chart). The shaded box is seen to represent 3.6 inches. Using the conversion, 3.6 in = 91.4 mm, so into this box is placed the figure 91.4.

In the same way each of the other boxes contains a number which is the conversion representing the inches corresponding to that box. The conversion values are correct to one place of decimals.

If we wish to convert 4.8 in to mm, we find the row with the value 4 in and then the column with the value 0.8. The intersection of the two, also shown as a shaded box, represents the millimetres, i.e.

4.8 in = 121.9 mm

(b) *CONVERSION OF METRES TO FEET*

From section 6.1(a), 1 m = 3.28 ft. Using the calculator we multiply each value of metres by the constant 3.28, using the same method as in section 6.2(a).

The conversion chart is made out in steps of 0.1 m as shown in Table 2.

Conversion chart, metres to feet, in steps of 0.1 m										
Metres	0.0	0.1	0.2	0.3	0.4	0.5	0.6	0.7	0.8	0.9
0	0.00	2.33	0.66	0.98	1.31	1.64	1.97	2.30	2.62	2.95
1	3.28	3.61	3.94	4.26	4.59	4.92	5.25	5.58	5.90	6.23
2	6.56	6.89	7.22	7.54	7.87	8.20	8.53	8.86	9.18	9.51
3	9.84	10.17	10.50	10.82	11.15	11.48	11.81	12.14	12.46	12.79
4	13.12	13.45	13.78	14.10	14.43	14.76	15.09	15.42	15.74	16.07

TABLE 2

(c) *CONVERSION OF FLUID OUNCES TO MILLILITRES*

The conversion factor is

1 fl oz = 28.4 ml

In Table 3 we take whole fl oz from 0 to 39, in steps of 1 fl oz.

Conversion of fluid ounces into ml, in steps of 1 fl oz										
fl oz	0	1	2	3	4	5	6	7	8	9
0	0	28.4	56.8	85.2	113.6	142.0	170.4	198.8	227.2	255.6
10	284.0	312.4	340.8	369.2	397.6	426.0	454.4	482.8	511.2	539.6
20 (1 pt)	568.0	596.4	624.8	653.2	681.6	710.0	738.8	766.8	795.2	823.6
30	852.0	880.4	908.8	937.2	965.6	994.0	1022.4	1050.8	1079.2	1107.6

TABLE 3

6.3 APPROXIMATE CONVERSIONS FOR USE WITH RECIPES

Most recipe books in Britain have traditionally been composed with weights in ounces (oz) and liquid measure in fluid ounces (fl oz). For example, the recipe for *Carbonnade de boeuf* is

$1\frac{1}{2}$ lb topside of beef
2 oz flour
2 oz dripping
12 oz sliced onions
$\frac{1}{2}$ pt brown ale
2 oz tomato purée
$1\frac{1}{2}$ pt brown stock

If we wish to use this recipe in the metric system we convert ounces to grams and fluid ounces to millilitres.

If we wish to use this recipe in the metric system we convert ounces to grams and fluid ounces to millilitres.

The exact conversions are:

1 oz = 28.4 g
1 fl oz = 28.4 ml

However, if we use approximations:

1 oz = 25 g
1 fl oz = 25 ml

although not exact, the proportions in the mixture will be the same, so that the receipe can be converted quite easily, that is, in the metric system we have:

Carbonnade de boeuf		
$1\frac{1}{2}$ lb	topside of beef	600 g
2 oz	flour	50 g
2 oz	dripping	50 g
12 oz	sliced onions	300 g
$\frac{1}{2}$ pt (10 fl oz)	brown ale	250 ml
2 oz	tomato purée	50 g
$1\frac{1}{2}$ pt (30 fl oz)	brown stock	750 ml

What does this conversion do in terms of the final dish? If we look at the reduction that takes place; we reduce 28.4 to 25 = a reduction of 3.4.

$$\text{Percentage reduction} = \frac{3.4}{28.3} \times 100 = 12\%$$

∴ The final dish will have a 12% reduction in the portion size if the approximate conversion factors are used.

The advantage of using these approximations, however, is that the recipe can be converted into metric so easily without resorting to the conversion charts.

EXAMPLE 1

Convert the following recipe into metric units.

Poulet sauté chasseur	
$\frac{1}{4}$ pt	vegetable oil
$2\frac{1}{2}$ lb	chicken cut for sauté
4 oz	onion
6 oz	button mushrooms
$\frac{1}{4}$ pt	dry white wine
1 pt	demi-glace
12 oz	tomatoes
	chopped parsley

Using the approximate conversions:

1 oz = 25 g
1 fl oz = 25 ml
(*Note*: 1 pt = 20 fl oz)

we have the following recipe in metric measurements.

Poulet sauté chasseur		
$\frac{1}{4}$ pt	vegetable oil	125 ml
$2\frac{1}{2}$ lb	chicken cut for sauté	1 kg*
4 oz	onion	100 g
6 oz	button mushrooms	150 g
$\frac{1}{4}$ pt	dry white wine	125 ml
1 pt	demi-glace	500 ml
12 oz	tomatoes	300 g
	chopped parsley	

$* \; 2\frac{1}{2}$ lb = 40 oz

40 oz = 25×40 g
= 1000 g
= 1 kg

6.4 DISTANCE CHARTS

Consider the distances in miles from Aberdeen to a number of towns listed below.

	Aberdeen to
Aberystwyth	445
Birmingham	420
Bristol	493
Cardiff	490
London	503

Consider the mileage from Aberystwyth now.

	Aberystwyth to
Birmingham	114
Bristol	125
Cardiff	105
London	211

These lists can now be combined on to a single chart.

	Aberystwyth	Aberdeen
		445
Birmingham	114	420
Bristol	125	493
Cardiff	105	490
London	211	503

The distances from Birmingham, Bristol and Cardiff are now considered.

Birmingham to	
Bristol	81
Cardiff	103
London	105

Bristol to	
Cardiff	45
London	115

Cardiff to	
London	167

These three sets of miles can now be further incorporated into a single diagram.

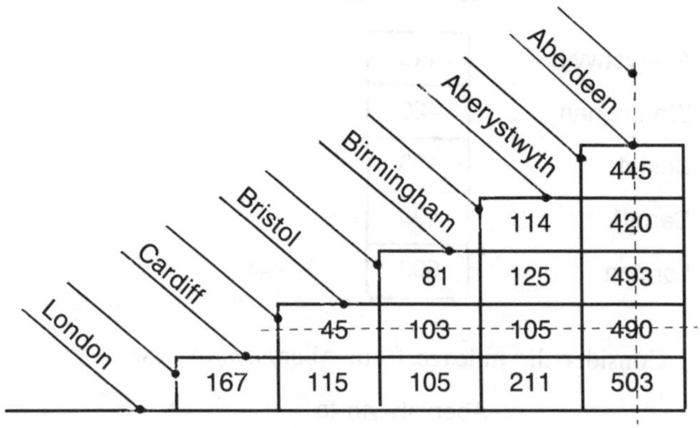

This forms the basis of a mileage chart. In order to find the distance between two towns, say Aberdeen and Cardiff, the point of intersection of a vertical line from Aberdeen and a horizontal line from Cardiff indicates the actual value at 490 miles. This type of chart is used by the motoring organizations, to provide a compact display of distances.

EXERCISES 6.1

1. Convert 0–10 inches to mm, in steps of 0.1 in, by extending Table 1.
2. Convert 0–10 m to feet, in steps of 0.1 m, by extending Table 2.
3. Convert 0–500 ml to fl oz, in steps of 10 ml.
4. Convert 0–50 miles to km, in steps of 1 mile.
5. Convert 0–50 pints to litres in steps of 1 pt.
6. Convert 0–5 lb to kg in steps of 1 oz.

7. Using the conversion rule in section 6.1(f), convert 30–180°F to °C, in steps of 5°C.

8. Using any road map, or any other source of information, draw up a distance chart for the following towns:
London, Cardiff, Glasgow, Liverpool, Plymouth, Manchester, Leicester.

9. Convert the following recipe to metric measure using approximate conversion factors:

Sauce Lomard	
$1\frac{1}{2}$	lobster
3 oz	butter
4 oz	onion
4 oz	carrot
2 oz	celery
$\frac{1}{8}$ pt	brandy
3 oz	flour
4 oz	tomato purée
$2\frac{1}{2}$ pt	fish stock
$\frac{1}{4}$ pt	dry white wine
	bouquet garne
$\frac{1}{2}$	crushed clove garlic
	salt

10. Convert the following recipe to British units, using the approximate conversion factors:

Moussaka	
50 g	onions
1	small clove garlic
25 g	butter
25 g	tomato purée
400 g	cooked mutton
125 ml	demi-glace
200 g	aubergine
200 g	tomatoes
	flour
62 ml	oil
25 g	breadcrumbs
25 g	grated parmesan cheese
	melted butter

6.5 GRAPHS TO REPRESENT CONVERSIONS

(a) *AXES*

The charts in the previous sections are made up of pairs of values, one value in each of two measuring systems. Between the pairs is a distinct relationship comprising the conversion value between the two sets of units.

Such relationships can be shown very clearly by diagrams. These diagrams are called **graphs**. Let us apply this idea to the conversion values of inches to millimetres, using the table below.

(in)	0	1	2	3	4	5	6	7	8	9
(mm)	0	25.4	50.8	76.2	101.6	127.0	152.4	177.8	203.2	228.6

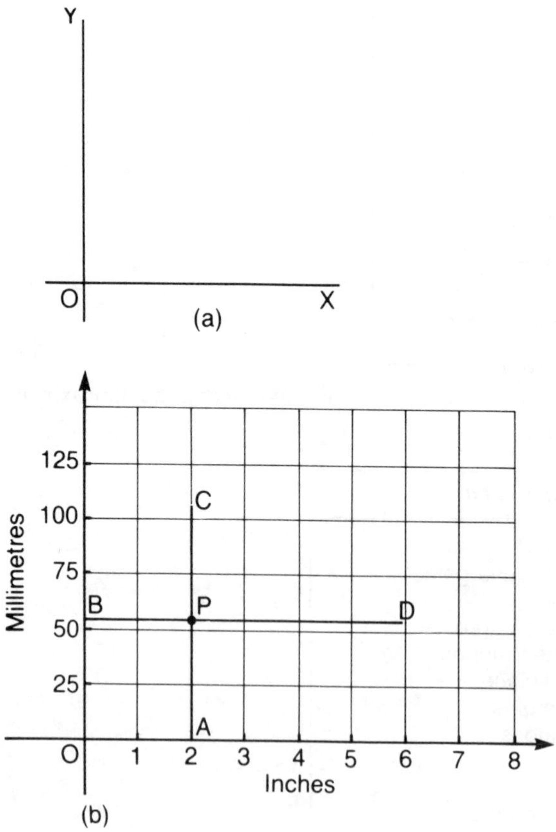

FIG. 6.1

These values are marked on two lines, normally at right angles to one another; called **axes**. In order to do so, of course, a suitable scale must be used. Such axes are shown as *OX* and *OY* in Fig. 6.1(a).

Since inches are converted to mm, the mm numbers are dependent upon the inches. These dependent values are usually marked along O*Y*. The inches values, which are independent, are marked along O*X*.

A suitable scale is chosen, similar to the markings on a rule. O*X* is marked in inches, with 1 inch per division, the millimetres are marked on O*Y*, with 25 mm per division, as shown in Fig. 6.1(b). Each division on an axis must always represent the same interval. For example, on the O*Y* axes, each division represents steps of 25 mm.

Consider now the pairs of values 2 in and 50.8 mm, marked at A and B in Fig. 6.1(b). From A we draw a vertical line AC and from B a horizontal line BD. The two lines intersect at P. The point P is now seen to pinpoint both the (in) and (mm) values.

In order to plot the values more easily, since they do not fall on the actual divisions, gridded paper is used as shown in Fig. 6.2. Each large division along O*X* is 1 in, so that each small division is 0.1 in. Along O*Y*, each large division is 25 mm, so that each small division is 2.5 mm.

The value pairs from the above tables are shown as P, Q, R, S, etc in Fig. 6.2. It is now found that a straight line can be drawn through all these points. This line OT is called a graph, which represents, diagrammatically, the relationship between inches and millimetres.

From the graph we can now determine other conversions than those in the table at the beginning of this section. For example, if we wish to find the number of millimetres corresponding to 5.4 in, we find 5.4 in along the O*X* axis at L. Draw LM vertically to meet the graph. Then draw MN horizontally to meet the O*Y* axis at M. The value at N is 135 mm, which is the number of millimetres equivalent to 5.4 in.

The conversion of millimetres to inches can be carried out in exactly the same way. To convert 110 mm to inches the lines DE and EF are drawn, where D marks 110 mm, which converts to 4.3 in.

(b) *SCALES*

In most instances, it is impossible to represent the actual values on an axis, for example, if one axis represented kilometres. We now allow a large division of the gridded paper to represent 1 km. Such a representation is called a **scale**. The important thing is to choose a scale for the large divisions which will allow easy or convenient subdivisions. As a working rule, we choose the large divisions to present 1 km, 2 km, 5 km, 10 km, 20 km, 50 km,

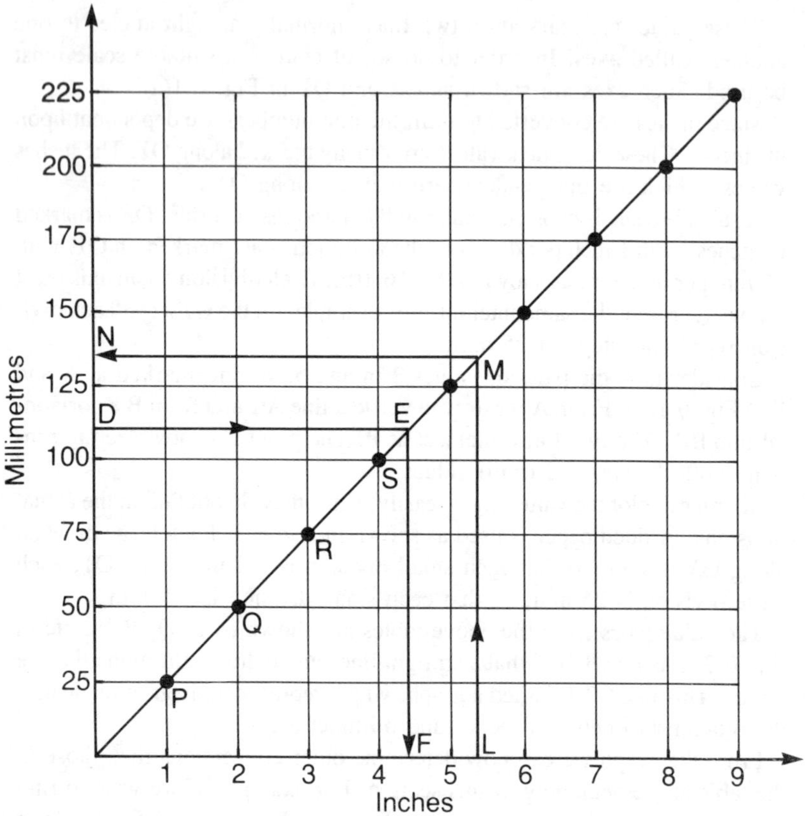

FIG. 6.2 Conversion of inches to mm

100 km depending on the actual magnitudes to be plotted. In this way, each subdivision represents a convenient value, as shown below.

Large division	Small division
1 km	0.1 km
2 km	0.2 km
5 km	0.5 km
10 km	1.0 km
20 km	2.0 km
50 km	5.0 km
100 km	10.0 km

EXAMPLE 2

Plot a graph of the conversion of miles to km, ranging from 0 to 90 miles.

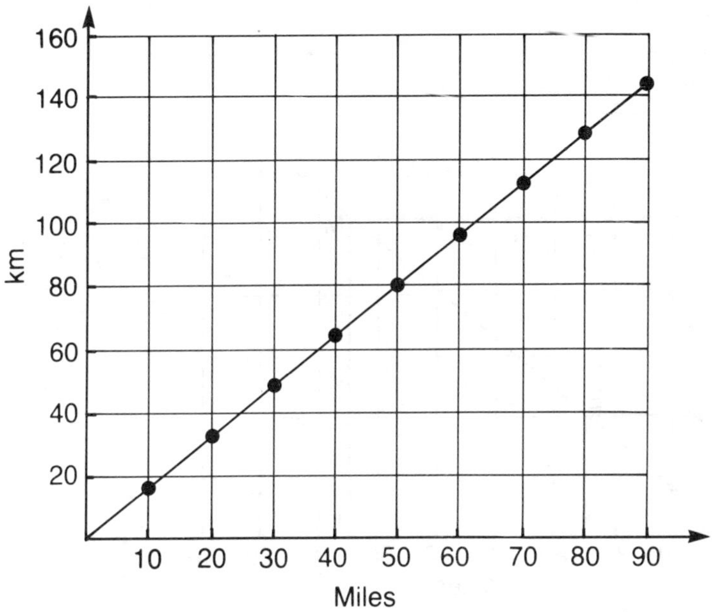

FIG. 6.3 Conversion of miles to km

First we produce the conversion table of miles to km, being the conversion factor in section 6.1(a).

miles	0	10	20	30	40	50	60	70	80	90
km	0	16.1	32.2	48.3	64.4	80.5	96.6	112.7	128.8	144.9

The graph is shown in Fig. 6.3.

In plotting such a graph as Fig. 6.3 all the points lie on a straight line. It is unnecessary to plot all the points; three points will do. Actually, two points are sufficient, but the third is a check to ensure that no errors are made. The data in Fig. 6.3 are now replotted, using three points only, in Fig. 6.4.

We can use Fig. 6.4 to convert miles to km, other than those in the table, and vice versa.

EXAMPLE 3

Using Fig. 6.4, convert: (i) 65 miles to km; (ii) 40 km to miles.

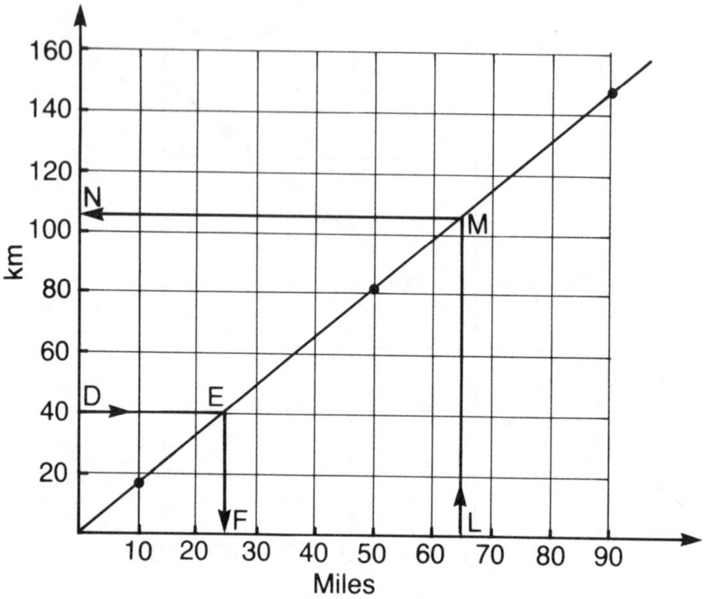

FIG. 6.4

(i) Let L be marked at 65 miles
Draw LM vertically to meet the graph
Draw MN horizontally to meet the km axis
The value corresponding to N is 105 km.

(ii) To convert 40 km to miles, we draw DE and EF, giving the value at F of 25 miles.

(c) *ACCURACY*

Because electronic calculators can handle figures so easily, the conversion tables can be assembled to very high accuracy. When graphs are used, however, the accuracy is limited by the subdivision. This explains why slightly different results were obtained when conversions were calculated using graphs.

(d) *CONVERSIONS OF °C TO °F*

Using just two points the conversion of temperature can be carried out quite easily.

Two conversion points are known:

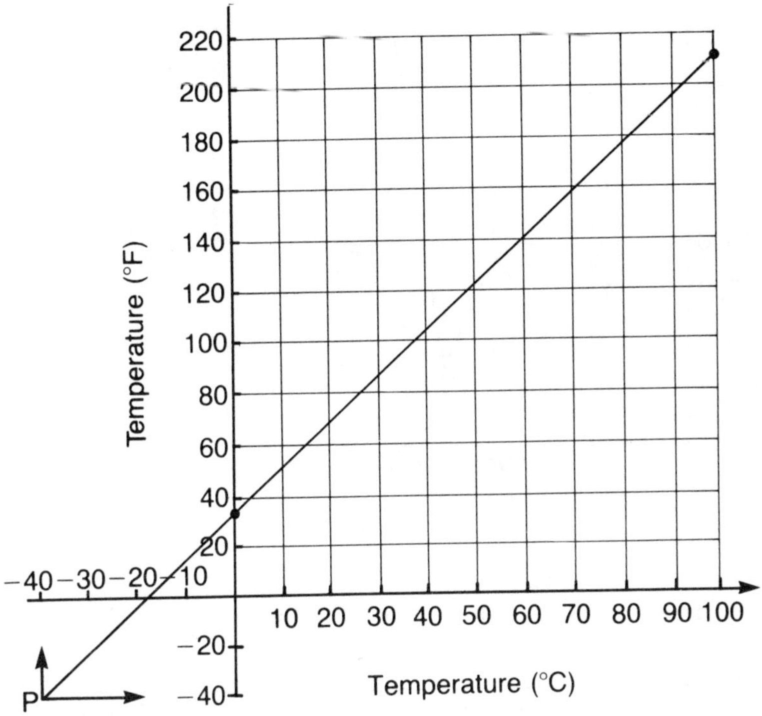

FIG. 6.5 Conversion of °C to °F

(i) at the freezing point of water, where 0°C ≡ 32°F
(ii) at the boiling point of water, where 100°C ≡ 212°F.

It will be sufficient in this case to use two points, because we know that both are correct from basic knowledge of the freezing and boiling points.

The graph is shown in Fig. 6.5.

It is interesting to note that if the graph is produced backwards, to the point P, it shows that at P both °C and °F have the same value of −40 that is,

$$-40°C = -40°F$$

6.6 GRAPHS OF PRACTICAL INFORMATION

In section 6.5 all the graphs were straight lines showing conversions from one set of units to another. The graphical approach, however, can be used to display many practical sets of data.

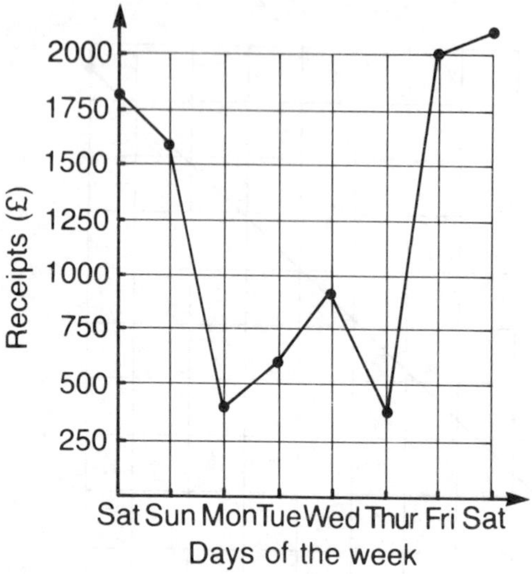

FIG. 6.6

Consider the following table, showing the daily receipts of a small restaurant over one whole week.

Sat	Sun	Mon	Tue	Wed	Thur	Fri	Sat
£1800	£1600	£400	£600	£900	£400	£2000	£2100

A graph plotted of the daily receipts is shown in Fig. 6.6. The scale chosen is £250 per large division.

The graph is no longer a straight line, since it shows the fluctuating demand on the restaurant. Although the 'takings' fluctuate, nevertheless, the graph shows a greater demand at the weekend, the demand coinciding with the general acceptance that the weekend is a non-working time, devoted to leisure and relaxation.

6.7 GRAPHS OF FIXED AND VARIABLE COSTS AND BREAKEVEN POINT

(a) *GRAPH OF FOOD COST AND FOOD SALES*

In Chapter 4 the term 'gross profit' was explained as the difference between food cost and food sales. We can plot graphs of these two transactions, against the number of meals.

Consider for example that the average cost of food materials in a college refectory is 50p per meal. The sale price is £1.25 per meal. We can now draw up a table of costs and sales and then plot a graph.

If the refectory sells 100 meals/week, the cost is 100 × 50p = £50. The income from these sales is £1.25 × 100 = £125. In the same way, we can calculate cost and sales for meals ranging up to 800 per week, as shown in the table below.

No. of meals sold/week	0	100	200	300	400	500	600	700	800
Food costs	0	£50	£100	£150	£200	£250	£300	£350	£400
Food sales	0	£125	£250	£375	£500	£625	£750	£875	£1000

A graph is drawn of food costs against meals sold and of food sales against meals sold, both drawn on the same axes as shown in Fig. 6.7.

The graph OA is drawn by plotting the number of meals per week against the food costs. Obviously, the more food that is prepared, the more the food costs are going to be. The graph OB is drawn by plotting the number of meals per week against the food sales.

The difference between the two graphs represents the gross profit. At a sales of 400 meals per week, the gross profit is given by LM, which is

£500 − £200 = £300

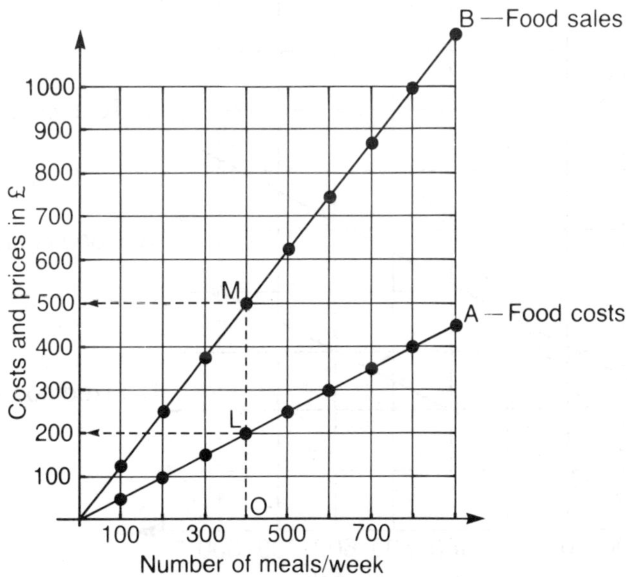

FIG. 6.7 Graph of food costs and food sales against number of meals/week

As more meals are sold, the gap between sales and costs becomes bigger, producing greater and greater profits.

(b) *GRAPH OF OVERHEADS AND FOOD COSTS*

If food costs in the preceding section were the only consideration, running a business would be very easy because a profit would be made whatever the number of meals sold. However, there are other costs, such as rates, salaries, which are the fixed costs that need to be taken into account.

Let us assume that in the example used a fixed cost of £200 per week existed. We can show this as in Fig. 6.8.

As before, OA is the graph of food costs. CD is the graph of fixed costs which exist whether 1 meal or 800 meals are sold. The total costs are given by the graph CE, which is drawn parallel to OA, but starting from C. The line CE represents the total cost, that is, food costs + fixed costs.

The actual income is again shown as OB. The graphs of sales and the

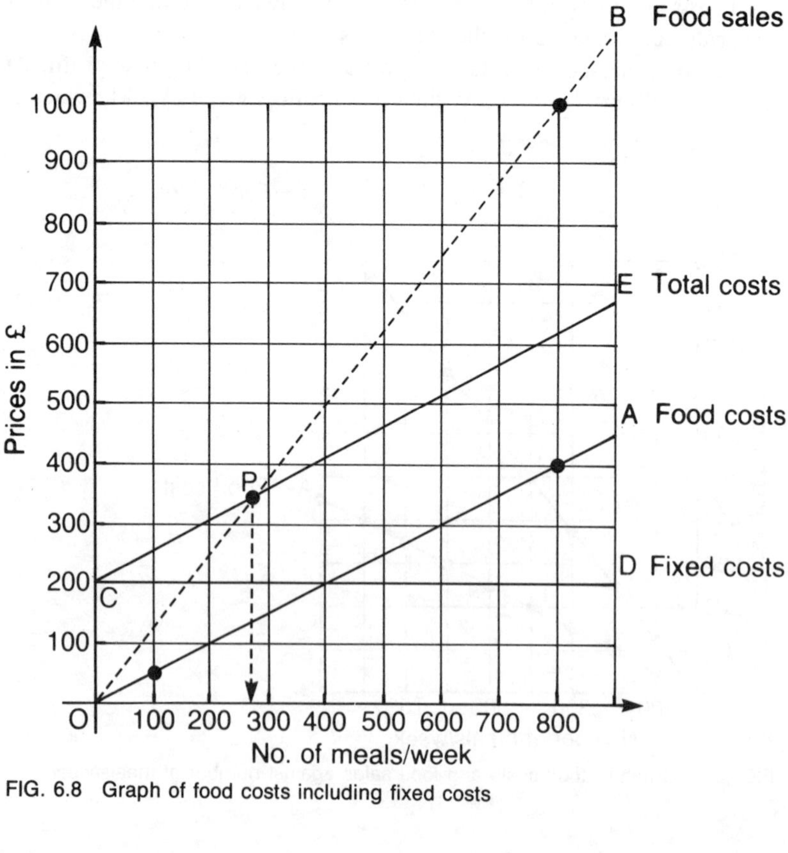

FIG. 6.8 Graph of food costs including fixed costs

total costs intersect at P. P is the **breakeven** point. This breakeven point is at a number of meals per week of 267. If the number of meals served per week is less than 267, the income from food sales is not enough to cover the fixed overhead costs.

EXERCISES 6.2

1. Plot a graph of the following table of figures, putting the number of meals on the horizontal axis and the total cost on the vertical axis.

Number of meals	0	10	20	30	40	50	60	70	80	90	100
Cost of meals	0	15	30	45	60	75	90	105	120	135	150

2. Plot a graph of the conversions of metres to feet over the range 0–60 m, using the conversion factor in section 6.1(a).

3. Plot a graph of the conversion of pints to litres as over the range 0–100 pints, using the conversion factor in section 6.1(c).

4. Draw a graph converting °F to °C over the range 20°F to 270°F.

5. Room bookings in a hotel over a 14-day period were:

Day	1	2	3	4	5	6	7	8	9	10	11	12	13	14
No. of rooms	20	23	60	63	49	30	22	6	30	49	52	56	21	16

Plot a graph of the information. Mark on the graph the weekends.

6. The booking of rooms were recorded as daily averages over a period from November to the end of January:

Week	1	2	3	4	5	6	7	8	9	10	11	12	13
Daily averages of rooms booked	45	43	49	10	10	49	64	65	30	31	42	41	40

Plot a graph of this information. Think of likely explanations for (i) drop in bookings on weeks 4 and 5, (ii) increase in bookings on weeks 7 and 8.

7. The proprietor of a restaurant estimates that the average amount paid by his customers for a meal excluding wine is £7.00. The cost of food material is £3.00 per meal.

(i) Plot a graph of total food costs ranging from 0 to 4000 meals served in the month.
(ii) If the overheads for the month are £8000, draw the graph of total costs.
(iii) Draw also the graph of total income and determine the breakeven point.

8. The works canteen is run by a firm of outside caterers. The maximum workforce is 250. On average the workforce spends £2.00 per meal. The food cost to the caterers is 90p on average per meal. Because of the nature of the work the numbers eating lunch fluctuates and the board of directors of the works has agreed to contribute £500 per week as guaranteed fixed income to the caterers.

Plot a graph of food costs against meals served per week, up to a maximum of 1500. If the fixed costs are £200 per week, draw the graph of total expenditure against the number of meals served.

Finally, draw a graph of food sales including the fixed income and find the breakeven point of meals served per week.

WORKED EXAMPLE 4

Upper Crust Food Enterprises is considering the acquisition of Ragamuffin's Restaurant. The purchase price is £100,000 and the directors of Upper Crust expect any new investment to show a return of **at least** 18%. The proprietor of the restaurant is able to provide the following information:

Total operating expenses £50,000 per annum
Wages (paid for 50 weeks) £1,000 per week
Restaurant opens 6 days a week for 50 weeks a year.

The average spent is £2.00 and the number of meals served averages 250 a day. The maximum number of meals it is possible to serve without structural alterations is estimated to be 350 a day. The gross food profit varies between 55% and 65%.

You are required to:

(i) calculate the target cash gross food profit necessary to meet the directors' requirements *(3 marks)*
(ii) construct a graph based on current trading conditions and state the number of meals which will break even at each level of gross profit to give the return on investment required *(7 marks)*
(*Hint — treat return on investment as a fixed cost*)
(iii) comment on the profitability of the current volume of trade and make a brief assessment of the potential profitability. Make recommendations on changes in trading policy which will ensure that the investment is successful. *(10 marks)*

(Final HCIMA)

(i) The returns on investment required to meet the directors' expectation is

18% of £100,000 = £18,000.

This can be regarded as a fixed cost in running the restaurant.

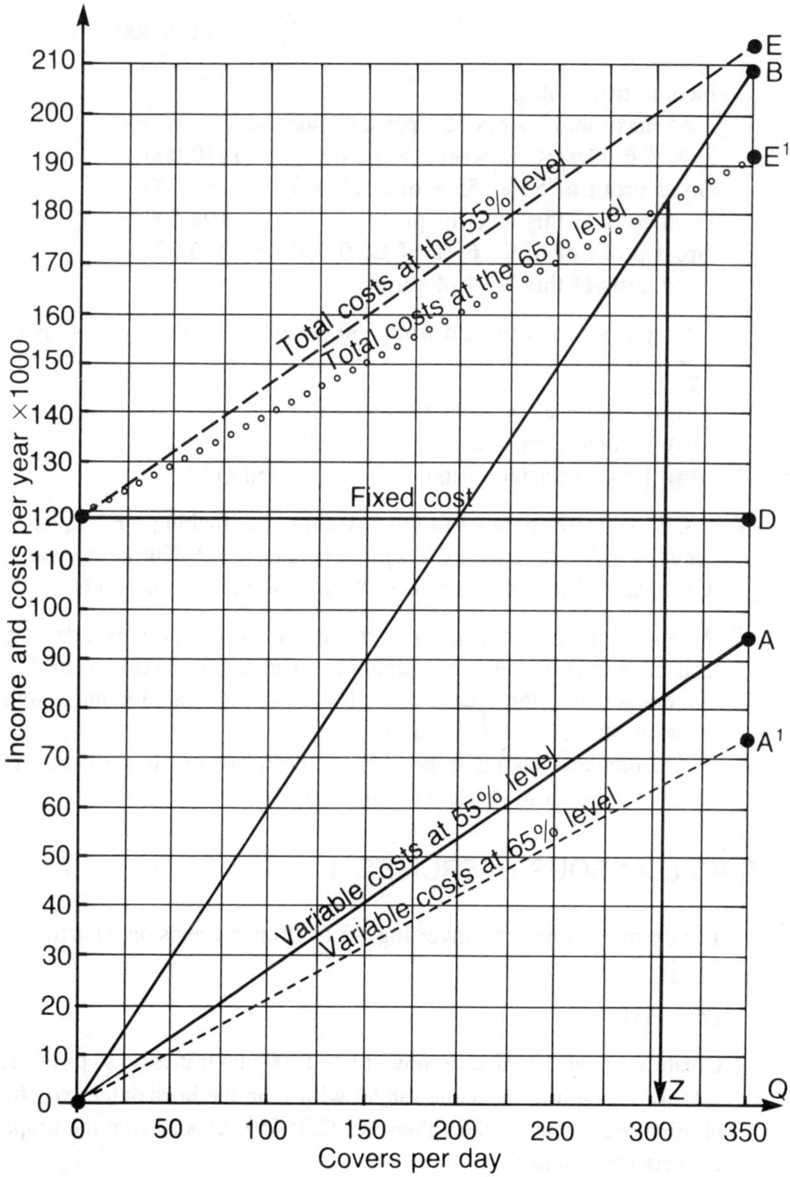

FIG. 6.9

(ii) Fixed costs:

Operating costs	=	£50,000
Wages for the year: 50 weeks × £1,000	=	£50,000
Return on investment	=	£18,000
		£118,000

Income from sales:

At maximum covers 350 per day, income in a year is

350 × 6 days × 50 weeks × £2.00	=	£210,000
Gross profit of 55%: 55% of £210,000	=	£115,500
Food costs at this level of profit	=	£94,500
Gross profit of 65%: 65% of £210,000	=	£136,500
Food costs at this level of profit	=	£73,500

Graphs are now plotted from 350 covers down to 0 covers, as in Fig. 6.9, of:

(i) Fixed costs = £118,000 — CD
(ii) Income from sales — OB
(iii) Food (variable costs) — OA and OA′

CE′ is drawn parallel to OA′, from C, by making DE′ equal in length to QA′. Likewise, CE is made parallel to OA. The points where CE′ and CE intersect the line OB are the breakeven points.

(iii) From the graph it is seen that at the average of 250 covers a day the return on investment is not attained. Even at the maximum of 350 covers per day, the breakeven will only occur when the maximum gross profit level is set at 65%

The breakeven point at the 65% level of gross profit is 305 covers per day, shown by the point X drawn to Z.

MISCELLANEOUS EXERCISES 6

1. The exchange rate for converting £ into French francs on a particular day is

£1 = 11f.

Create a conversion table to convert £0–£100 into francs, in steps of £1.

Draw a graph in the same range, with £ on the horizontal axis, by plotting four points at the values £0, £20, £50, £100. From the graph convert 450f. into £.

2. A car is stated to be able to do 40 miles to the gallon of petrol. Draw a graph of miles against gallons over the range 0 to 15 gallons.

3. The following figures represent total sales and total costs in a large works canteen:

Week no.	1	2	3	4	5	6	7	8	9	10
Total sales	£ 1700	1700	1600	1800	1700	2000	1800	2000	1850	1850
Total costs	£ 1500	1400	1400	1600	1700	1700	1700	1750	2000	1800

Plot these values on a graph, using a continuous line for sales and a broken line for costs. Use the graph to answer the following questions:

(a) Which week was most profitable?
(b) How much profit was made in that week?
(c) In which week did sales equal costs?
(d) In which week was there a net loss?
(e) How much was this loss?

(Oxford)

4. Given the information that 5.5 lbs = 2500 grams, construct a graph which can be used to convert the quantities in the following recipe from Imperial to metric units.

Choose a suitable scale and state this on your graph.

Blanquette de veau

Veal	$5\frac{1}{2}$ lbs
Onions	3 lbs 10 oz
Carrots	8 oz
Mushrooms	2 lbs
Flour	4 oz
Margarine	3 oz

(Oxford)

5. The standing charge on electricity is £8.00 per quarter. The cost of electricity is 5.3p per unit used. Plot a graph of total cost of electricity over a range 0 to 2000 units.
(*Note*: Standing charge is added to cost of units).

6. Sales in a restaurant, over a year period, are given in the table below:

Month	Jan	Feb	Mar	Apr	May	June	July	Aug	Sept	Oct	Nov	Dec
Sales in £	2200	1700	1700	1800	1500	1200	2000	2000	1800	1800	1900	2000

Plot a graph of sales per month.
From your graph state when you think the restaurant was closed for two weeks.

(Blackpool & Fylde)

7. Bacteria in a food dish increase at the rate of 11% per day. From a sample on a given day, it was estimated that the dish contained 1000 bacteria. Calculate the bacteria present at the end of each day over a two-week period. Place your results into a table:

Day	1	2	3	4	5	6	7	8	9	10	11	12	13
No. of bacteria													

and plot a graph of the results.
Make any deduction from the graph.

(GLOSCAT)

8. The Fine Meals Co. is a catering operation which has the following selling price and cost structure:
Food cost per cover £1.50
Fixed costs per week £1800.00
There are no other costs
Selling price per cover £4.00
Sales histories show the range of demand per week as between 800 and 1000 covers.

Draw graphs of food sales, food costs and fixed costs against covers in the range 800 to 1000 covers per week.

From your graph determine the number of covers per week to make a profit of £400 per week.

7

CALCULATIONS IN THE CATERING INDUSTRY

7.1 INTRODUCTION

One simple calculation which is used all the time in the catering industry is finding the number of portions that may be obtained from a given quantity of basic ingredients. The basis of the calculations is simply that of dividing a given quantity of food into an appropriate number of portions.

A typical calculation is how many 4 oz slices of meat may be obtained from a piece of meat weighing 80 oz:

Number of pieces $= 80 \div 4 = 20$

7.2 NUMBER OF PORTIONS

We shall do the calculations in both the British and the metric systems.

(a) *USING BRITISH WEIGHTS*

How many 4 oz portions may be obtained from a piece of meat weighing 12 lb?

This is the same calculation as in section 7.1, but first of all, the 12 lb must be expressed in ounces.

Now $\quad 1 \text{ lb} = 16 \text{ oz}$
$\quad 12 \text{ lb} = 12 \times 16 \text{ oz}$

$$\text{Number of 4 oz portions} = \frac{12 \times 16}{4} = 48$$

EXAMPLE 1

How many 2 oz pats of butter may be obtained from 5 lb?

$$5 \text{ lb} = 5 \times 16 \text{ oz}$$

$$\text{Number of 2 oz pats} = \frac{5 \times 16}{2}$$

$$= 40$$

EXAMPLE 2

How many glasses of spirit, each of $\frac{5}{6}$ fl oz, may be obtained from a bottle of volume $26\frac{2}{3}$ fl oz?

In this case both are already in fl oz, but the quantities involve fractions.

Number of glasses is the number of $\frac{5}{6}$ units in $26\frac{2}{3}$

$$= 26\tfrac{2}{3} \div \tfrac{5}{6}$$

$$= \frac{\overset{16}{80}}{3} \times \frac{\overset{2}{6}}{5}$$

$$= 32$$

(b) *USING METRIC WEIGHTS*

This is a similar calculation, but now portions are expressed in grams and the total quantity in kg.

Note, 1 kg = 1000 g

EXAMPLE 3

Find how many 100 g cutlets may be obtained from a fish weighing 8 kg.

First of all, the 8 kg must be expressed in grams.

$$8 \text{ kg} = 8 \times 1000 \text{ g}$$

$$\text{Number of 100 g cutlets} = \frac{8 \times 1000}{100}$$

$$= 80$$

7.3 YIELD AND PORTION SIZE

When raw foods are purchased, such as meat, fish and vegetables, they contain parts which are unusable. These parts are bones, peel, stalks, heads and tails of fish, etc., which are trimmed off to obtain the usable food. This usable food is called the **yield**.

For example, a short sirloin of beef weighs 15 kg. It contains $6\frac{3}{4}$ kg of bone and trimmed fat. The usable meat is therefore

$$15 \text{ kg} - 6\tfrac{3}{4} = 8\tfrac{1}{4} \text{ kg}$$

$$= 8.25 \text{ kg or } 8250 \text{ g}$$

This is the yield.

$$\text{The percentage yield is } \frac{\text{usable meat}}{\text{original weight}} \times 100$$

$$= \frac{8.25}{15} \times 100$$

$$= 55\%$$

From this usable amount, we can calculate the number of portions that may be obtained from it.

Number of 100 g portions (approx. 4 oz) $= \frac{8250}{100} = 82.5$

Number of 125 g portions (approx. 5 oz) $= \frac{8250}{125} = 66$

Number of 150 g portions (approx. 6 oz) $= \frac{8250}{150} = 55$

In addition to trimming losses there are preparation losses in the cooking. These are losses caused by the evaporation of water, fat and juices.

EXAMPLE 4

A piece of meat weighing 25 kg loses 40% in trimming and cooking. Calculate the weight of usable meat and the number of whole portions that will be obtained from it, if the portions weigh

(i) 100 g
(ii) 125 g
(iii) 150 g.

Weight of meat when purchased = 25 kg
Total percentage loss = 40%
Actual weight loss on trimming and cooking = 40% of 25 kg
= 10 kg
Usable meat = 15 kg
= 15 000 g

Number of 100 g portions $= \frac{15000}{100}$

$= 150$

Number of 125 g portions $= \frac{15000}{125}$

$= 120$

Number of 150 g portions $= \frac{15000}{150}$

$= 100$

EXAMPLE 5

A salmon weighing 12 kg lost 2.4 kg on trimming head and tail and cleaning, and a further 1.6 kg when grilled. Calculate the total percentage loss and the number of portions of weight 200 g which could be served from it.

Total loss from trimming and cooking $= 2.4 + 1.6$
$= 4$ kg

Percentage loss $= \frac{4}{12} \times 100$

$= 33\frac{1}{3}\%$

Usable fish $= 8$ kg

Number of portions from usable fish $= \frac{8000}{200}$

$= 40$

EXAMPLE 6

A pineapple weighs 3 lb. After trimming and removing the core, $2\frac{1}{4}$ lb of fruit remains. Calculate:

(i) the percentage loss
(ii) the number of 3 oz portions obtained from the usable fruit.

Loss $= 3 - 2\frac{1}{4}$ lb

$= \frac{3}{4}$ lb

Percentage loss $= \dfrac{\frac{3}{4}}{3} \times 100$

$= \frac{3}{4} \times \frac{1}{3} \times \frac{100}{1}$

$= 25\%$

To find number of portions, convert $2\frac{1}{4}$ lb to oz.

$2\frac{1}{4}$ lb $= 2\frac{1}{4} \times 16$ oz

Number of 3 oz pieces $= 2\frac{1}{4} \times 16 \div 3$

$= \frac{9}{4} \times \frac{16}{1} \times \frac{1}{3}$

$= 12$

EXERCISES 7.1

1. Find how many
(i) 4 oz pieces,
(ii) 6 oz pieces,
may be obtained from a piece of meat weighing 24 lb.

2. Calculate how many
(i) 100 g
(ii) 150 g
(iii) 200 g
cutlets may be obtained from a fish weighing 14 kg.

3. How many 8 oz slices of steak may be obtained from a piece of meat weighing 14 lb?

4. A bowl of paté weighs 7 kg including the weight of the bowl. The empty bowl weighs 2.5 kg. Find how many 50 g portions may be obtained.

5. 56 lb of potatoes were purchased. Some potatoes were rotten. After peeling 49 lb of sound potatoes remained. Determine the percentage loss and the number of 5 oz portions that may be obtained.

6. 5 kg of fresh peas in pod, when shelled yield $1\frac{1}{4}$ kg of peas. Calculate the percentage of peas in the total. How many portions may be obtained if each portion is 50 g?

7. 25 lb of fresh tomatoes are to be used in a side salad. On opening the case 8% are found to be rotten. Calculate the weight of usable tomatoes. Calculate the number of 6 oz portions that can be obtained.

8. Carrots are served in 100 g portions. What weight of carrots is required for a dinner for 30 people? If 10% is lost in trimming and cleaning, calculate the weight of fresh carrots required.

7.4 COSTING OF PORTIONS

The management of any establishment determines the percentage profit to be made on each item of food sold in the restaurant. Having established the percentage profit, the price of a portion to the customer has to be calculated. The following steps have to be followed:

Step 1 Specify the gross percentage profit on the items of food.
Step 2 Determine the selling price in relation to the buying price.

Step 3 Calculate the usable weight of the food.
Step 4 Specify the portion size and calculate the number of portions.
Step 5 Calculate the price of the portions to the customer.

This procedure is illustrated below:
A piece of beef weighing 28 kg is purchased for £96. On trimming and cooking it loses 12 kg. The gross percentage profit is specified as 60%. Calculate:

(i) the actual selling price of the beef
(ii) the price of 200 g portions to the customer.

Step 1	Gross percentage profit	= 60%
Step 2	Cost price	= £96
	If the selling price be 100%	
	Cost price is 100 − 60	= 40%
	Now 40%	= £96
	By proportion, 100%	$= £\frac{96}{40} \times 100$
	Selling price	= £240
Step 3	Usable weight	= 28 − 12 kg
		= 16 kg
		= 16 × 1000 g
Step 4	Portion size	= 200 g
	Number of portions obtained	$= \dfrac{16 \times 1000}{200}$
		= 80
Step 5	Price per portion	$= £\frac{240}{80}$
		= £3.00

Therefore, in the meat course served to the customer meat portions will contribute £3.00 to the total price.

EXAMPLE 7

A whole salmon weighing 15 lb is purchased for £33. After trimming, cleaning and grilling, 12 lb of usable fish remains. The fish is served on the bone in 6 oz cutlets. Calculate:

(i) the number of portions that may be obtained,
(ii) the price of each cutlet to the customer if the gross percentage profit is set at 45%.

Step 1 Gross percentage profit = 45%

Step 2 Cost price $= £33$

Let the selling price be 100%

Cost price is therefore $100 - 45 = 55\%$

Therefore $55\% = £33$

By proportion, $100\% = \frac{33}{55} \times 100$

$= £60$

That is, selling price $= £60$

Step 3 Usable weight $= 12$ lb

$= 12 \times 16$ oz

Step 4 Portion size $= 6$ oz

$$\text{Number of portions} = \frac{12 \times 16}{6}$$

$= 32$

Step 5 Price per portion $= £\ \frac{60}{32}$

$= £1.87$

EXERCISES 7.2

1. A fillet of beef weighs 6 lb. Trimming and cooking account for 2 lb. The cost price of the fillet is £5.50 per lb with £3 allowed on a discount. Calculate the selling price if a gross profit of 60% is required. Hence, determine the price per portion to the customer if each portion is

(i) 4 oz

(ii) 8 oz.

2. Grilled gammon steak is the main course for a lunch menu. The gammon is purchased in a piece weighing 18 lb costing 60p per lb. Calculate:

(i) the selling price if a gross percentage profit of 70% is to be made

(ii) the number of cooked portions of size 6 oz that can be obtained, assuming 10% loss on trimming and 20% loss on cooking

(iii) the cost per portion to the customer.

3. Pineapples weighing 30 kg were purchased, costing £1.60 per kg. Loss of weight from trimming was 20%. If the gross profit was expected to be 70% find the total selling price. Find too, the number of portions, if each portion is 150 g, and the cost per portion to the customer.

4. A salmon weighing 25 lb is purchased for £2 per lb. In cleaning and cooking 25% is lost. If the gross profit is to be 60% find the total price at which it is sold to the customer.

If the salmon is served in 4 oz pieces, calculate the number of portions obtained and the cost per portion to the customer.

5. Paring waste of apples is about 25%. A box of apples weighing 32 kg is purchased for £24 (less 5% discount). Calculate the cost price per kg of edible fruit. Determine also the selling price per kg if a gross profit of 55% is planned.

6. The waste in trimming and preparation of cauliflower is 65%. A box of cauliflowers is purchased for £22. It weighs 45 lb and contains 12 cauliflowers, graded to approximately the same weights. Calculate the weight of edible vegetables in the box.

In preparing *Chou-fleur au gratin*, each cauliflower is cut into quarters. What is the weight of each portion served? If an overall gross profit of 45% is required find the cost of each portion to the customer.

7.5 COSTING OF DISHES — COSTING PLANS

When a dish for a course on a menu is prepared by the chef, it will contain, or use, many more than one ingredient. Accurate pricing is an essential part of the system of control and is essential in arriving at the price of the course placed on the menu.

The following illustrates a method of finding the cost of a dish. A clear layout of the costing exercise is essential if errors are to be avoided.

(a) *RECIPE*

The recipe for *Sauté de boeuf Strogonoff* for 4 portions is:

1 lb of fillet of beef
2 oz of butter
$1\frac{1}{2}$ oz of onions
measure of brandy
$\frac{1}{4}$ pt of cream
$\frac{1}{2}$ lemon juice
1 bunch of parsley
2 cloves

(b) *PRICES*

These prices are merely illustrative to show the method of carrying out the exercise. Prices vary so much that they are out of date very quickly.

beef at £3.50 per lb
butter at 72p per lb
shallots at 64p per lb
wine at £2.40 per bottle
cream at £1.20 per pt
lemon 20p each
parsley 26p per bunch

(c) *COSTING PLAN*

Quantity	Cost of purchase	Actual ingredient cost		
1 lb of beef	£3.50 per lb			£3.50
2 oz of butter	72p per lb	$\frac{2}{16}$ of 72p	=	0.09
$1\frac{1}{2}$ oz of onions	64p per lb	$\frac{3}{32}$ of 64p	=	0.06
measure of brandy	£2.40 per bottle	$\frac{1}{4}$ of £2.40	=	0.60
$\frac{1}{4}$ pt of cream	£1.20 per pt	$\frac{1}{4}$ of £1.20	=	0.30
$\frac{1}{2}$ lemon	20p each	$\frac{1}{2}$ of 20p	=	0.10
1 bunch of parsley	26p per bunch		=	0.26
	Total cost			£4.91
	Cost of each portion			£1.23

(d) *PRICE TO THE CUSTOMER*

Assuming now, that a profit of 70% is required, then we can calculate the price to the customer.

Selling price = 100%
Profit = 70%
Cost price = 100 − 70 = 30%
Therefore 30% = £1.23

$$100\% = \frac{1.23}{30} \times 100 = £4.07$$

EXERCISES 7.3

Using the following pricing list, draw up a costing plan for each of the recipes below.

almonds (ground)	£1.10/$\frac{1}{4}$ lb
apricot glaze	£0.65/lb
bread	£0.40/lb
bacon	£0.80/lb
brandy	£0.60/measure
butter	£1.40/lb
chicken	£0.60/lb untrimmed (75% wastage on trimming)
cream	£1.60/pt
carrots	£0.40/lb
demi-glaze	£0.90/pt
eggs	£0.60/doz
flour	£0.48/pack of 3 lb
fat	£0.40/lb
ham	£0.60/$\frac{1}{4}$ lb
jam (raspberry)	£0.65/lb
milk	£0.40/pt
onions	£0.30/lb
pork	£1.20/lb
parsley	£0.30/oz
pistachio nuts	£1.10/$\frac{1}{4}$ lb
sugar	£0.38/lb
tongue	£0.60/$\frac{1}{4}$ lb
tomato purée	£0.20/$\frac{1}{4}$ lb
veal	£1.80/lb
veal stock	£0.40/pt

1. *Epaule de veau farcie* (4 portions)
8 oz carrots
4 oz lard
8 lb shoulder of veal
2 oz tomato purée
2 pt veal stock
$1\frac{1}{2}$ pt demi-glaze

8 oz parsley
20 oz suet
1 oz thyme
2 oz butter
16 oz breadcrumbs
4 oz onions
Calculate the cost price per portion and the sales price to the customer if the gross profit is to be 50%.

2. *Galantine de volaille* (8 portions)
$4\frac{1}{2}$ lb chicken
$\frac{1}{4}$ pt double cream
4 eggs
$\frac{1}{2}$ oz pistachio nuts
2 oz ham
2 oz tongue
Calculate the cost price per portion and the sales price to the customer if the gross profit is 70%.

3. *Bakewell tart* (8 portions)
8 oz sugar
8 oz flour
$1\frac{1}{2}$ oz icing sugar
2 oz raspberry jam
2 oz apricot glaze
4 oz butter
2 oz ground almonds
2 eggs
4 oz caster sugar
2 oz flour
Find the cost of preparing the tart for 80 people and the total bill for the tart to make a gross profit of 60%.

7.6 BAR SALES

In selling spirits over the bar, a bottle is expected to yield a given number of units, via the optics.

In the British units, which still operate, a bottle of spirit — brandy, whisky, gin — has a volume of $26\frac{2}{3}$ fl oz.

(i) Spirits are served in tots of $\frac{1}{6}$ of a gill ('6 out' as called in the trade).

$$1 \text{ pt} = 20 \text{ fl oz}$$

$$1 \text{ gill} = \tfrac{1}{4} \text{ pt} = 5 \text{ fl oz}$$

$$\therefore \tfrac{1}{6} \text{ gill} = \tfrac{5}{6} \text{ fl oz}$$

$$\text{Number of tots of spirit in a bottle} = 26\tfrac{2}{3} \div \tfrac{5}{6} = \tfrac{80}{3} \times \tfrac{6}{5} = 32$$

(ii) Sherry and port are served in tots of $\frac{1}{3}$ of a gill ('3 out').

$$\tfrac{1}{3} \text{ gill} = \tfrac{5}{3} \text{ fl oz}$$

$$\text{Number of tots in a bottle} = 26\tfrac{2}{3} \div \tfrac{5}{3} = \tfrac{80}{3} \times \tfrac{3}{5} = 16$$

(iii) Liqueurs are served in tots of $\frac{1}{5}$ gill ('5 out').

$$\tfrac{1}{5} \text{ gill} = 1 \text{ fl oz}$$

$$\text{Number of tots in a bottle} = 26\tfrac{2}{3} \div 1 = 26$$

EXAMPLE 8

From one bottle of whisky, 17 tots have been served at 65p a tot. What should the takings be and what is the value of the spirit remaining in the bottle?

Takings of the whisky sold = 17 × 65p
= £11.05

The number of tots remaining in the bottle is 32 − 17
= 15

Value of the remaining spirit = 15 × 65
= £9.75

EXAMPLE 9

The sales of gin in the bar of a hotel one evening are as follows:

Opening stock	Issued from the cellar	Closing stock
$2\frac{3}{32}$ bottles	10 bottles	$5\frac{7}{32}$ bottles

Find the cash in the till from the gin sales that evening if each tot of gin is priced at 55p.

$$\begin{aligned}\text{Gin sold during the evening} &= \text{Opening stock} + \text{issued} - \text{closing stock}\\ &= 2\tfrac{3}{32} + 10 - 5\tfrac{7}{32}\\ &= 7 + \tfrac{3}{32} - \tfrac{7}{32}\\ &= 6 + \tfrac{32}{32} + \tfrac{3}{32} - \tfrac{7}{32}\\ &= 6\tfrac{28}{32}\end{aligned}$$

In one whole bottle there are 32 tots, so that the takings on 6 bottle are

$6 \times 32 \times 55\text{p} = £105.60$

$\frac{28}{32}$ means 28 tots, the price for which is

$28 \times 55 = £15.40$

Total sales of gin = £121.00

EXERCISES 7.4

1. A bottle of brandy has served 19 tots at 70p a tot. How much should the total income be?

2. The bar sales of spirits one evening in a hotel bar were:

   Whisky $3\frac{18}{32}$ bottles at 60p a tot

   Gin $4\frac{9}{32}$ bottles at 64p a tot

   Brandy $3\frac{19}{32}$ bottles at 70p a tot

   Vodka $1\frac{9}{32}$ bottles at 61p a tot

   Find the total sales of spirit.

3. A bottle of whisky is purchased for £9.00 a bottle from the wholesaler. Over the bar the tot of whisky is priced at 70p. Find the gross percentage profit.

4. Calculate the sales of the following:
Whisky 1 bottle and 14 measures at 70p a measure
Sherry 2 bottles and 5 measures at 45p a measure
Liqueur 12 measures at 60p a measure

5. A bottle of whisky is purchased for £8.96. If a gross profit of 65% is to be made, calculate the selling price and the price of each tot for which it must be sold.

6. A bottle of sherry is purchased for £2.16 and sold to make a gross profit of 70%. Find the price of each tot sold in the bar.

7.7 OTHER CALCULATIONS

(a) *ELECTRICITY*

In using electricity we pay so many pence per unit. The unit is called the kilowatt-hour. A bar of an electric fire is rated as one kilowatt, written as 1 kW. If this is left on for 1 hour it uses one kilowatt-hour (1 kW h) of electric energy. Similarly, we have the rating of other appliances:

	Rating	Units of energy (kW h) used in 1 hour
Electric iron	1 kW	1
Kettle	2 kW	2
Domestic oven	6 kW	6
Refrigerator	$\frac{1}{4}$ kW	$\frac{1}{4}$

EXAMPLE 10
A refrigerator runs on average 3 hours out of a 24-hour day. If the motor is rated at $\frac{1}{4}$ kW h, find how many units of electrical energy are used in 1 week. Calculate how much this costs if 1 kW h is charged at 5p, to the nearest p.

Rating $= \frac{1}{4}$ kW

Amount used in 3 hours $= \frac{1}{4} \times 3$ kW h

$= \frac{3}{4}$ kW h

Amount used in 1 week $= \frac{3}{4} \times 7$

$$= \frac{21}{4}$$

$$= 5\tfrac{1}{4}\,\text{kW h}$$

Total cost $= 5\frac{1}{4} \times 5$

$$= 5.25 \times 5$$
$$= 26.25$$
$$= 26\text{p}$$

EXAMPLE 11

Work out the following electricity bill:

Meter readings				
Previous	Present	Unit charge	VAT	£
28,524	30 858	? @ 5.32p	0	
		Quarterly standing charge		7.54
		Total due		

The number of units used in the quarter is

30858 − 28524 = 2334

Cost is

2334 @ 5.32p = £124.16

Total bill is £124.16 + £7.54 = £131.70

(b) *GAS*

Gas is paid for using the same type of calculation, the bill being arrived at from the total cost of units of gas used plus the standing charge.

(c) *THE TELEPHONE BILL*

A telephone bill looks something similar to the following:

	Rental	From	To	Quarterly rate	£
PART A		1.2.86	30.4.86	16.45	16.45

	Metered units	Date	Meter reading	£
PART B		24.10 30.1	007358 008235	
	Number of units used		877	
	Units at 5.00p		877	43.85
	Total charge			£60.30
	VAT at 15%			9.05
	Total payable			£69.35

The first part of the bill, labelled A here, gives the quarterly rent of the telephone system.

Part B shows how many units are used in the quarter, which is the difference, 877, between the two meter readings at the beginning and end of the quarter. In this particular bill, the charge is 5.00p per unit.

The total charge is the unit costs + rental.

Added to this is 15% VAT; making a total payable of £69.35.

(d) *MILEAGE*

Two types of calculations arise: one to do with miles per gallon, the other miles per hour. In calculations dealing with miles per gallon, we are concerned with petrol consumption and how far a car can travel on a given quantity. Petrol consumption is rated in miles per gallon, for example, a car travels 35 miles to the gallon. For every gallon of petrol used, the car travels 35 miles.

In the second case, we are dealing with distances travelled in a given time.

EXAMPLE 12

If a car does 35 miles to the gallon, find:

(i) how many miles it will travel with 6 gallons in the tank

(ii) how many gallons it requires to travel 385 miles.

(i) 1 gallon will allow the car to travel 35 miles, so that on 6 gallons, it travels

$$6 \times 35 = 210 \text{ miles}$$

(ii) To travel 385 miles, we want to find how many 35s there are in 385, that is:

$$\frac{385}{35} = 11$$

Thus 11 gallons of petrol are required.

Speeds of cars are expressed in miles per hour, for example, 30 miles per hour means that a car will travel:

30 miles in 1 h
60 miles in 2 h
90 miles in 3 h
15 miles in $\frac{1}{2}$ h
45 miles in $1\frac{1}{2}$ h

and so on.

To find how far the car travels in a given time, we multiply speed by time, that is:

$$\boxed{\text{distance} = \text{speed} \times \text{time}}$$

If we wish to find how long it will take to make a journey, we find how many 30 miles there are in the distance travelled. For example, if we wish to find how long it takes to travel 240 miles, at a speed of 30 miles per hour, we find how many 30s there are in 240, that is:

$\frac{240}{30} = 8$

∴ It takes 8 hours to travel 240 miles.

The rule for finding the number of hours is:

$$\boxed{\text{time} = \frac{\text{distance}}{\text{speed}}}$$

EXAMPLE 13

A car travels at a speed of 45 miles per hour. Find how long it will take to travel 180 miles and the distance travelled in $5\frac{1}{2}$ hours.

(i) Number of hours is the number of 45s in 180

$$= \frac{180}{45}$$
$$= 4$$

∴ Time taken $= 4$ hours

(ii) Distance travelled is the speed × time

$$= 45 \times 5\tfrac{1}{2}$$
$$= 247.5 \text{ miles}$$

EXERCISES 7.5

1. An electric fire is switched on from 6 p.m. to 11 p.m. each evening for 13 weeks. It is rated at 2 kW. Find the total cost of running the fire if each unit costs 5p.

2. An electric cooking range in the kitchen is rated at 12 kW and is used on average for 2 hours each lunchtime and 4 hours each evening, seven days a week. Calculate the cost of cooking for a week if each unit costs 4.8p a unit.

3. Complete the following electricity bill:

Meter readings				
Previous	Present	Unit charge	VAT	£
30 8584	32 719	? @ 5.41p	0	
	Quarterly standing charge			8.12
	Total due			

4. Work out the following telephone bill:

	Rental	From	To	Quarterly rate	£
PART A		2.2.86	31.4.86	£17.56	£17.56
	Metered units		Date	Meter reading	
PART B			25.10 31.1	006172 006921	
	Number of units used				
	Cost of calls @ 5.00p				
	Total charge VAT at 15%				
	Total charge for the quarter				

5. A car is quoted as being able to travel 43 miles to the gallon. Calculate the number of gallons required to travel 516 miles.

How much does the journey cost if 1 gallon costs £1.80? Calculate the VAT included in this petrol bill at 15%.

MISCELLANEOUS EXERCISES 7

1. A recipe for a cake lists the following ingredients:

	Weight	Cost
Flour	250 g	14p/kg
Sugar	200 g	24p/kg
Margarine	200 g	44p/kg
Other ingredients	100 g	25p

Calculate:

(i) the weight of the cake, assuming a 20% loss of weight during cooking
(ii) the total cost of the cake
(iii) what is the total cost of the cake to a customer in order to make 70% profit on the cost of materials?

(Oxford C of FE)

2. The liquor control records from a banquet bar produced the following figures:

Item	Issued	Returned	Wholesale cost per dozen bottles	Price
Dry Sherry	6	$\frac{14}{16}$	£22.08	25p per measure
Medium Sherry	6	$1\frac{8}{16}$	£19.44	25p per measure
Sweet Sherry	6	$3\frac{1}{16}$	£21.89	25p per measure
Bin No. 36	12	2	£18.40	£3.10 per bottle
Bin No. 97	12	6	£23.92	£4.20 per bottle
Bin No. 48	12	1	£27.11	£4.50 per bottle
Brandy	4	$1\frac{25}{32}$	£64.30	45p per measure
Port	4	$\frac{12}{16}$	£43.00	45p per measure

Compile the account for presentation to the customer.

(HCIMA intermediate)

3. A cash banquet bar gives the following consumption figures:

		Wholesale Cost
Gin	$3\frac{1}{4}$ bottles	£4.14 per bottle
Whisky	$3\frac{3}{4}$ bottles	£4.29 per bottle
Rum	$\frac{7}{8}$ bottles	£4.45 per bottle
Brandy	$1\frac{1}{2}$ bottles	£5.94 per bottle
Light Ale	62 bottles	£1.39 per dozen
Lager	104 bottles	£1.44 per dozen
Tonic water	126 bottles	86 per 6 doz case
Ginger Ale	154 bottles	86 per 6 doz case
Lemonade	31 bottles	£1.18 per dozen
Soda	22 bottles	86 per 6 doz case
Pepsi-Cola	59 bottles	£1.18 per dozen

The bar is required to return 40% gross profit.

(a) Calculate a selling price per measure (6 out per spirit measure) for each commodity.

(b) Prepare the account for presentation to the customer.

(HCIMA intermediate)

4. The main course planned for a lunch menu for 80 covers is grilled gammon steak. The portion size required is 6 oz cooked meat. Labour costs are 35% of sales.

Grade A Gammon is purchased in 14 lb pieces @ 45p lb.

Preparation loss has been calculated at 17%, and cooking loss at 22%.

Calculate:

(a) the size of the order of gammon.

(b) the number of portions produced.

(c) the cost per portion.

(d) the 'after wage %' with a selling price of 85p per portion.

(e) a selling price to give 62% gross profit.

(HCIMA intermediate)

5. A short sirloin of beef weighs 20 lb and costs 64p per lb. It produces

1 lb 8 oz fillet
7 lb 8 oz contrafillet
3 lb flank
8 lb bones, fat and preparation loss (assumed to have no value).

(a) Calculate the cost per portion of:

Fillet	6 oz
Contrafillet	8 oz and
Flank	7 oz

(b) Show how you would calculate a selling price for each cut of meat and give an example.

(HCIMA intermediate)

6. A dispense bar produced the following figures for the last week in May:

Item	Opening stock	Issues from cellar	Closing stock	Selling prices
Gin	$7\frac{6}{32}$	24	$5\frac{4}{32}$	35
Whisky	$9\frac{1}{32}$	24	$11\frac{27}{32}$	35
Brandy	$3\frac{30}{32}$	12	$5\frac{9}{32}$	35

Calculate the total income that should be achieved.

(HCIMA intermediate)

8
AREAS AND VOLUMES

8.1 MEASUREMENT OF AREA

Area is a measure of surface coverage. The basic units of measurement are:

(i) metric system: $1\text{ m} \times 1\text{ m} = 1\text{ m}^2$
(ii) British system: $1\text{ ft} \times 1\text{ ft} = 1\text{ ft}^2$ (Fig. 8.1).

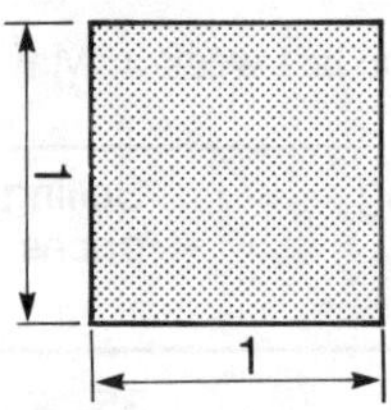

FIG. 8.1 The basic unit of area 1 m × 1 m or 1 ft × 1 ft

(a) *AREA OF A SQUARE*

In Fig. 8.2 a square is drawn 3 m by 3 m. If we break up the area into basic units of 1 m by 1 m, as shown by the dotted lines, we have 9 units of area, that is,

$A = 9\text{ m}^2$

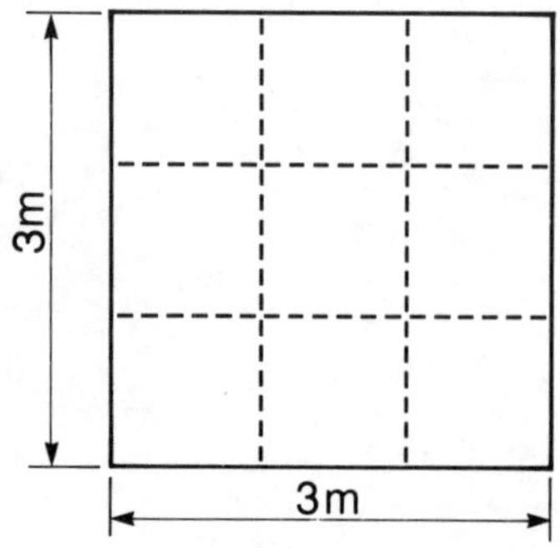

FIG. 8.2 Area of a square 3 m × 3 m

This may be obtained by multiplying the length by the breadth, that is,

$$\begin{aligned} A &= 3 \times 3 \\ &= 9 \text{ m}^2 \end{aligned}$$

In general, we have a square, shown in Fig. 8.3, in which the square has sides of length L. Again the area of the square is

$$\begin{aligned} A &= L \times L \\ &= L^2 \end{aligned}$$

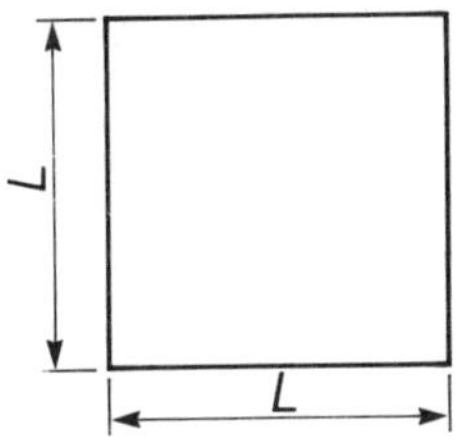

FIG. 8.3 Area of a square of side L: $A = L \times L$

It is far easier to carry out the calculation of multiplying the two sides together than trying to subdivide a surface into basic units. The calculation can then be extended to include fractional values, as shown in Example 1.

EXAMPLE 1

Calculate the area of a square whose sides are 5.1 m.

$$\begin{aligned} \text{Area} &= L \times L \\ &= 5.1 \times 5.1 \\ &= 26.01 \text{ m}^2 \\ &= 26.0 \quad \text{corrected to one place of decimals} \end{aligned}$$

(b) *AREA OF A RECTANGLE*

In Fig. 8.4 a rectangle of length 4 m and breadth 3 m is illustrated. Splitting it up into 1 m squares produces 12 squares, i.e. it has an area of 12 m^2. This is obtained by multiplying the length by the breadth, i.e.

$$\begin{aligned} \text{Area} &= 4 \times 3 \\ &= 12 \text{ m}^2 \end{aligned}$$

In general, an area of a rectangle is given by the formula

$$A = L \times B$$

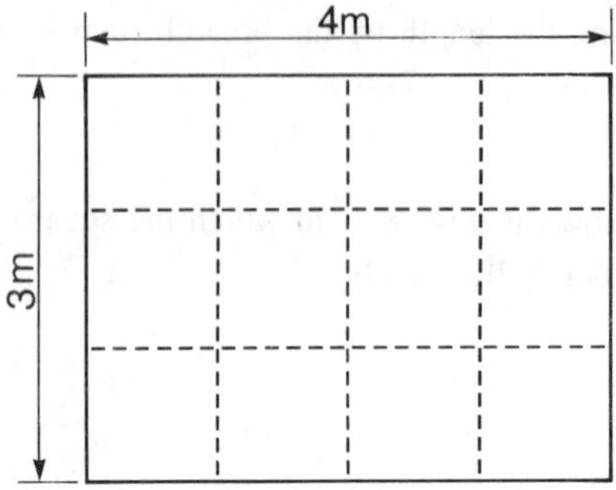

FIG. 8.4 Area of a rectangle of sides 4 m × 3 m

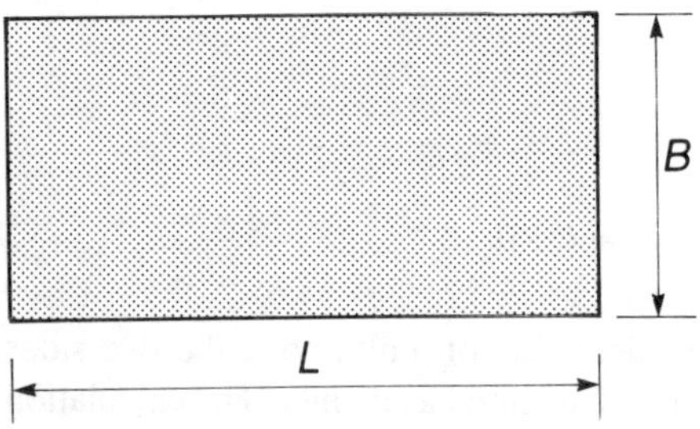

FIG. 8.5 Area of a rectangle of sides $L \times B$

where L is the length and B is the breadth of the rectangle, as shown in Fig. 8.5.

(c) *CALCULATION OF MORE COMPLEX AREAS*

Figure 8.6 shows a diagram of a floor of a room, 12 m × 10 m, covered by a carpet, 10 m by 8 m. The area of the uncovered surrounding border is required.

The easiest method of finding the area of the surround is to calculate the area of the floor and then the area of the carpet and subtract the area of the carpet from the floor area:

Area of floor	$= 12 \times 10 =$	$120\ \text{m}^2$
Area of carpet	$= 10 \times 8 =$	$80\ \text{m}^2$
Area of surround	$=$	$40\ \text{m}^2$

EXAMPLE 2

A table in a restaurant is $3\frac{1}{3}$ m × $1\frac{1}{3}$ m. A tablecloth is required which hangs $\frac{1}{3}$ m all the way round. Calculate the area of the tablecloth.

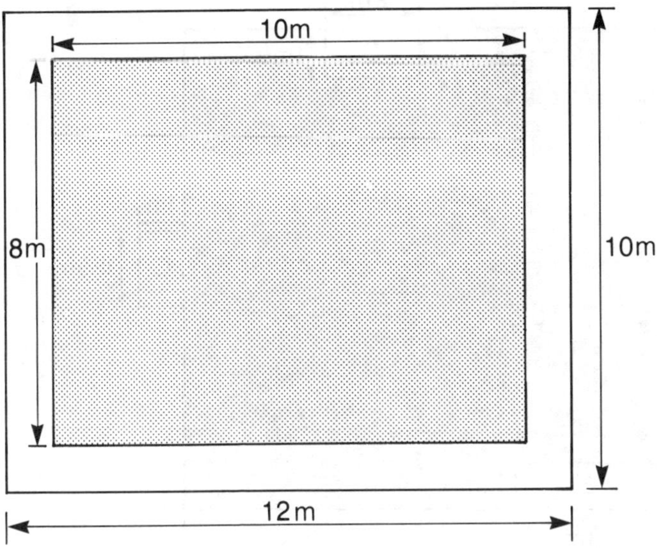

FIG. 8.6 Floor of a room, carpeted as shown by the shaded area

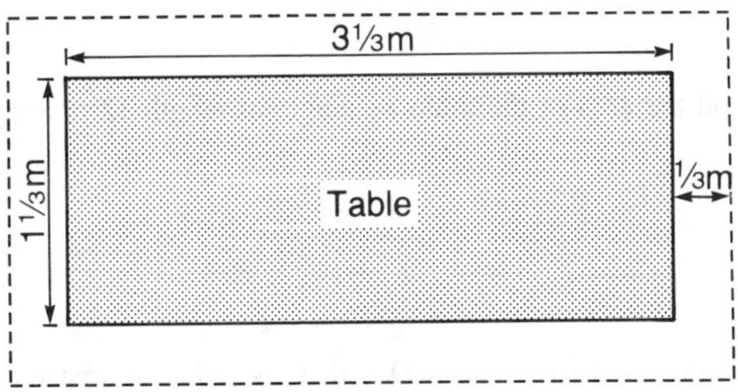

FIG. 8.7 Area of a tablecloth

As shown in Fig. 8.7,

Length of the tablecloth will be $3\frac{1}{3} + \frac{1}{3} + \frac{1}{3} = 4$ m

Breadth of the tablecloth will be $1\frac{1}{3} + \frac{1}{3} + \frac{1}{3} = 2$ m

Area of the tablecloth will be $4 \times 2 = 8$ m^2

EXAMPLE 3

A floor in a hotel has a peculiar shape shown in Fig. 8.8. Calculate its area using the dimensions shown.

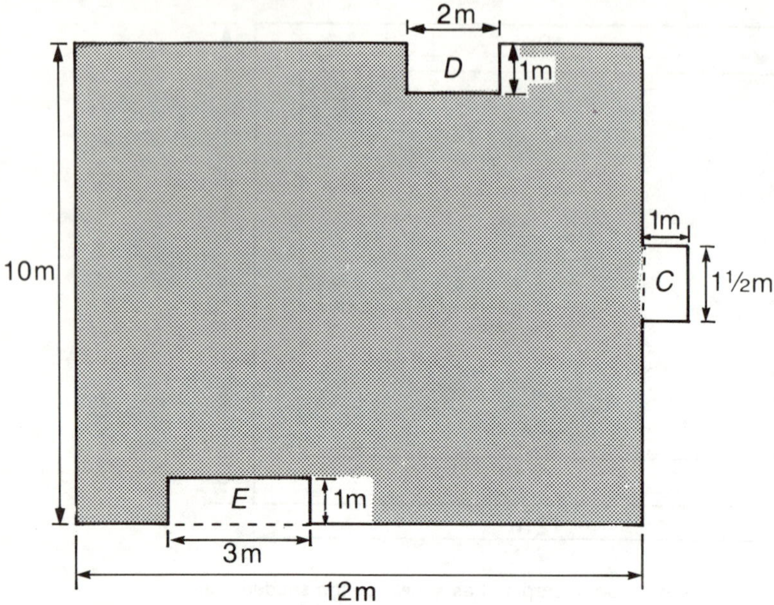

FIG. 8.8 Floor of a hotel room

The method is to find the area of the main rectangle of length 12 m and breadth 10 m, that is

$$\begin{aligned}\text{Area} &= L \times B \\ &= 12 \times 10 \\ &= 120 \text{ m}^2\end{aligned}$$

From this area we subtract the two areas D and E and add the area C.

$$\begin{aligned}\text{Area } D &= 2 \times 1 \\ &= 2 \text{ m}^2\end{aligned}$$

$$\begin{aligned}\text{Area } E &= 3 \times 1 \\ &= 3 \text{ m}^2\end{aligned}$$

$$\begin{aligned}\text{Area } C &= 1 \times 1\tfrac{1}{2} \\ &= 1\tfrac{1}{2} \text{ m}^2\end{aligned}$$

$$\begin{aligned}\therefore \text{ Area of floor} &= 120 - 2 - 3 + 1\tfrac{1}{2} \\ &= 115 + 1\tfrac{1}{2} \\ &= 116\tfrac{1}{2} \text{ m}^2\end{aligned}$$

EXERCISES 8.1

1. Find the areas of the surfaces having the following dimensions:
(a) 4 m × 4 m
(b) 6 m × 6 m
(c) 7 m × 13 m
(d) 8 m × 4 m
(e) $3\frac{1}{2}$ m × $4\frac{2}{5}$ m
(f) 5.1 m × 6.3 m
(g) 7.2 m × 8.3 m

2. Find the areas of the shaded shapes in Fig. 8.9.

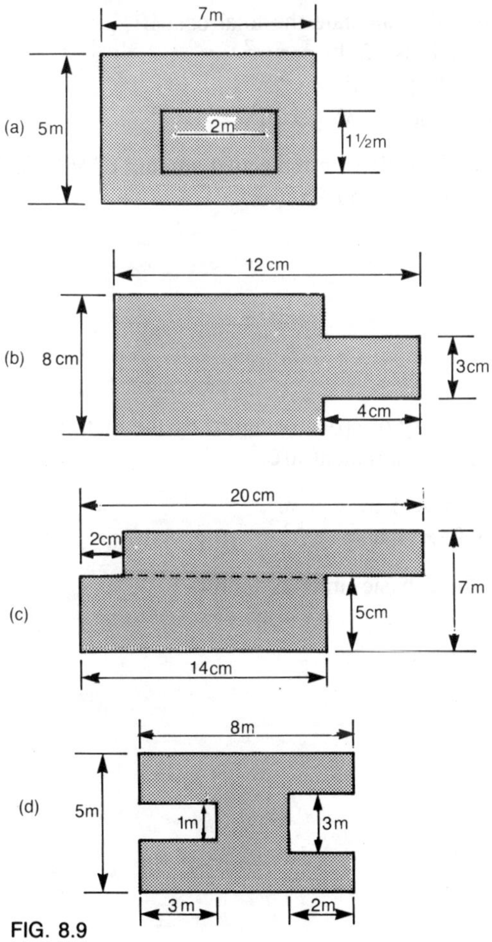

FIG. 8.9

3. A long top table in a formal dinner is 12 m long; the table has a width of 2 m. A tablecloth is to have $\frac{1}{2}$ m overhang all the way round. Calculate the area of the tablecloth.

If the table is to be covered by tablecloths of 3 m in length and an overlap of $\frac{1}{2}$ m is required where they join, how many tablecloths are required for the top table?

4. A lounge of a hotel is 13.1 m × 10.4 m. It is going to be carpeted leaving 1.2 m of a surround. Find the area of the carpet.

5. A room is 5 m × 4.5 m × 2.5 m. Calculate the total surface area of the walls.

The room is to be refurbished. Calculate the total cost if:

(i) wallpaper for the walls is £2.40 per m^2
(ii) ceiling tiles cost £1.20 per m^2
(iii) carpet for the floor costs £6.35 per m^2.

These costs include 15% VAT. Calculate the actual amount of VAT in order to make the return to Customs and Excise.

8.2 VOLUMES

Volume is a measure of the capacity or space occupied by the material of a solid or liquid. The units of measurement are:

(i) metric system: 1 m × 1 m × 1 m = 1 m^3
(ii) British system: 1 ft × 1 ft × 1 ft = 1 ft^3 (Fig. 8.10).

There are subdivisions of these basic units as follows:

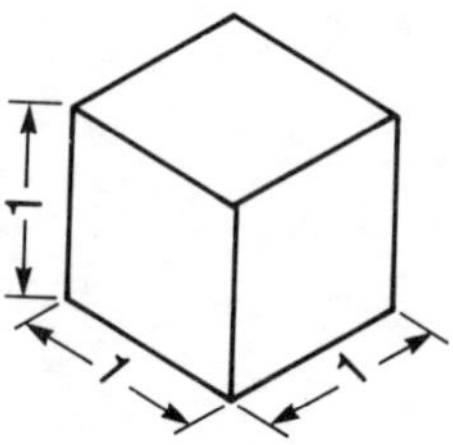

FIG. 8.10 The basic unit of volume, 1 m × 1 m × 1 m or 1 ft × 1 ft × 1 ft

(i) *Metric system.* A convenient measure of volume is 1 cm × 1 cm × 1 cm = 1 cm^3. The cubic centimetre (cm^3) is also written as a millilitre (ml). Another convenient measure is the litre (ℓ).

The relationships between them are:

1 m^3 = 1000 ℓ
1 ℓ = 1000 ml

Another unit which is used in liquid measure is the centilitre (cl), where 10 ml = 1 cl.

(ii) *British system.* The units of volume measurement are:
cubic foot (ft^3), gallon (gal), pint (pt),
fluid ounce (fl oz), gill.

The relationships between them are:

1 ft^3 = $6\frac{1}{4}$ gal
1 gal = 8 pt
1 pt = 4 gill or 20 fl oz
1 gill = 5 fl oz

EXAMPLE 4

Convert:

(i) 6.2 m^3 to ℓ
(ii) 1.4 ℓ to ml
(iii) 4261 ml to ℓ
(iv) 721 ℓ to m^3

(i) 6.2 m^3 = 6.2 × 1000 ℓ
= 6200 ℓ
(ii) 1.4 ℓ = 1.4 × 1000 ml
= 1400 ml
(iii) 4261 ml = 4261 ÷ 1000 ℓ
= 4.261 ℓ
(iv) 721 ℓ = 721 ÷ 1000 m^4
= 0.721 m^3

To convert m^3 to ℓ, or ℓ to ml, we multiply by 1000.
To convert ml to ℓ, or ℓ to m^3, we divide by 1000.

EXAMPLE 5

Convert:

(i) 4 gal to pt
(ii) 6 pt to fl oz
(iii) 100 fl oz to pt

(i) 4 gal $= 4 \times 8$ pt
$= 32$ pt
(ii) 6 pt $= 6 \times 20$ fl oz
$= 120$ fl oz
(iii) 100 fl oz $= 100 \div 20$ pt
$= 5$ pt

(a) *VOLUME OF A CUBE*

In Fig. 8.11 a cube is drawn 3 m × 3 m × 3 m. If we break up the cube into basic units of 1 m × 1 m × 1 m, as shown by the dotted lines, we have 27 units of volume, that is,

$$V = 27 \text{ m}^3$$

This may be obtained by multiplying length × breadth × height, that is,

$$\begin{aligned} V &= 3 \times 3 \times 3 \\ &= 27 \text{ m}^3 \end{aligned}$$

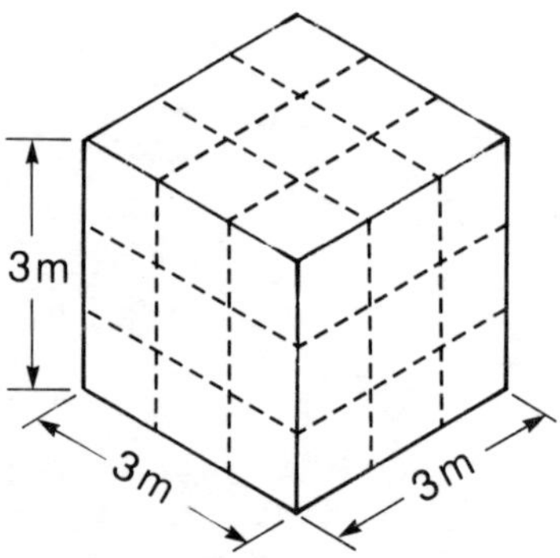

FIG. 8.11 Volume of a cube 3 m × 3 m × 3 m = 27 m^3

(b) *VOLUME OF A RECTANGULAR PRISM*

The volume of the rectangular prism of length L, breadth B and height H (as shown in Fig. 8.12) is

$$V = L \times B \times H$$

EXAMPLE 6
Find the volume of a rectangular prism of length 5 m, breadth 3 m and height 2 m.

$$\begin{aligned} \text{Volume} &= L \times B \times H \\ &= 5 \times 3 \times 2 \\ &= 30 \text{ m}^3 \end{aligned}$$

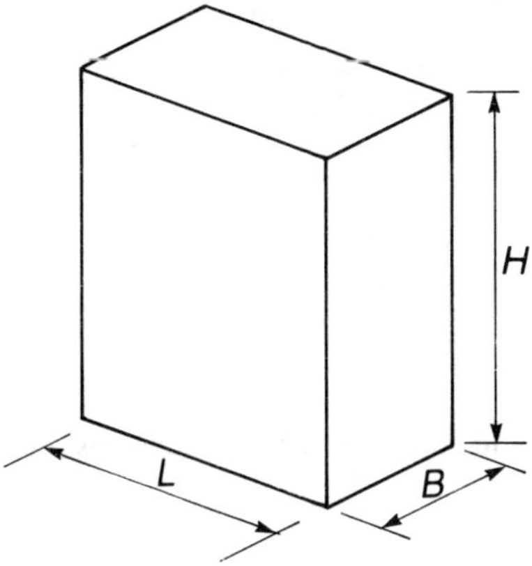

FIG. 8.12 Volume = $L \times B \times H$

EXAMPLE 7

A three-tiered wedding cake, each of height 10 cm, has the following square dimensions:

Bottom tier 40 cm × 40 cm
Middle tier 25 cm × 25 cm
Top tier 15 cm × 15 cm.

Calculate the total volume of the cake.

The cake is shown in Fig. 8.13.

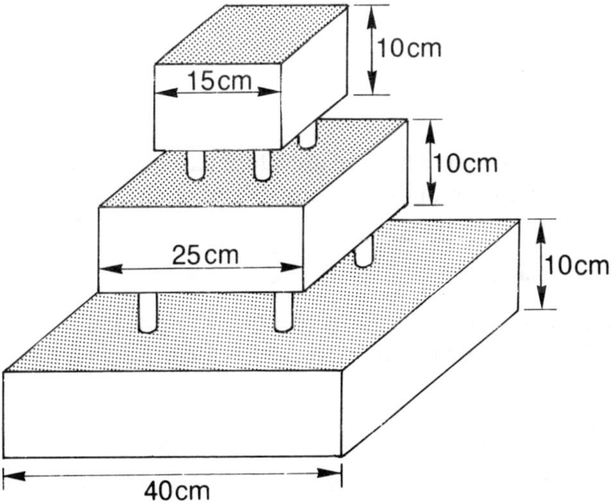

FIG. 8.13

Volume of bottom tier is $40 \times 40 \times 10$	=	16 000 ml
Volume of middle tier is $25 \times 25 \times 10$	=	6 250 ml
Volume of top tier is $15 \times 15 \times 10$	=	2 250 ml
Total volume		24,500 ml

EXERCISES 8.2

1. Find the volumes of the figures with the following dimensions and convert them into litres:
   (a) 10 cm × 3 cm × 15 cm
   (b) 6 cm × 6 cm × 6 cm
   (c) 9 cm × 6 cm × 11 cm
   (d) 9.3 cm × 8.4 cm × 7.7 cm
   (e) 6.2 cm × 13.1 cm × 8.6 cm

2. A room is 5 m long, 4.5 m wide and 2.5 m high. Calculate the volume of the room. If an extractor fan removes 5 m^3 of air per minute, how long does it take to have a complete change of air in the room?

8.3 CIRCLES, SPHERES AND CYLINDERS AND CONES — USE OF FORMULAE

The areas and volumes of circular figures are more complicated and to calculate them requires the use of formulae.

(a) *USE OF FORMULAE*

In order to calculate areas or volumes, we use a formula, which is a shorthand way of writing out a set of instructions. We have already used an example of this technique so far:

Volume of a rectangular prism $= L \times B \times H$

which is an instruction to multiply length by breadth by height. It will apply to any set of figures. If the length is 10 cm, breadth is 8 cm, height is 6 cm, then

$L = 10$ cm, $B = 8$ cm, $H = 6$ cm

so that we can **substitute** these figures into the formula, that is

$$V = 10 \times 8 \times 6$$
$$= 480 \text{ ml.}$$

We apply this technique to circular and spherical figures. The × signs are omitted generally — they are understood to be present, so that one may write

$$V = LBH$$

(b) *THE CIRCLE*

Parts of a circle are shown in Fig. 8.14.

One of the surprising things about a circle is that the distance around the edge, called the **circumference** (see Fig. 8.14) divided by the diameter always gives the same result, that is

$$\frac{\text{circumference}}{\text{diameter}} = 3.142$$

This number 3.142 is therefore a very special number and is given a special symbol; π.

We use two formulae with the circle:

Circumference = 2 × π × radius: $C = 2\pi r$
Area = π × radius × radius: $A = \pi r^2$

EXAMPLE 8
Find the circumference and area of a circle of radius 8 cm, correct to 2 places of decimals. Take $\pi = 3.142$.

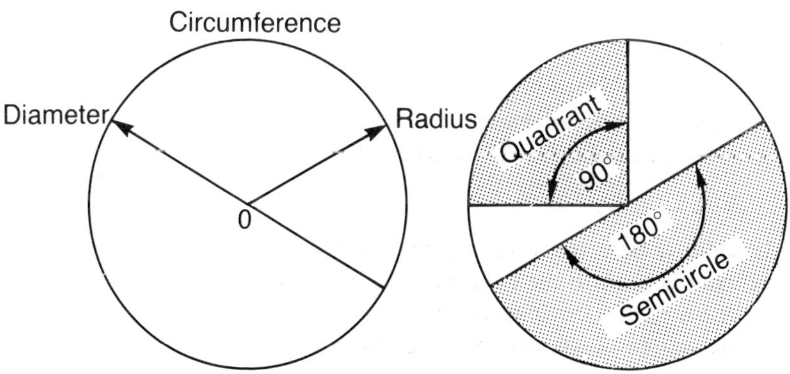

FIG. 8.14 Parts of a circle

(i) Circumference

$$C = 2\pi r$$
$$= 2 \times 3.142 \times 8$$
$$= 50.265$$
$$= 50.26 \text{ cm} \quad \text{correct to 2 decimal places}$$

(ii) Area

$$A = \pi r^2$$
$$= 3.142 \times 8 \times 8$$
$$= 201.062$$
$$= 201.06 \text{ cm}^2 \quad \text{correct to 2 decimal places}$$

(c) *THE CYLINDER*

The cylinder is shown in Fig. 8.15.

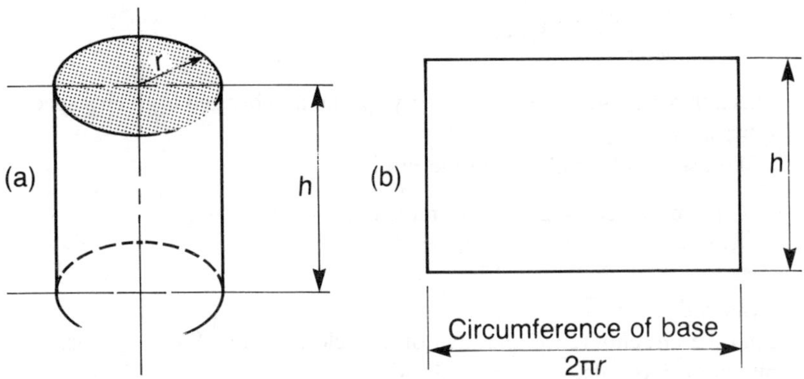

(a) The cylinder (b) The side of the cylinder opened out

FIG. 8.15

The surface area around the side, called the curved surface area, is seen to be the circumference × height, that is,

$$A = C \times h = 2\pi rh.$$

In addition there are two areas, top and bottom, which are circles, if the cylinder is solid. If the cylinder is hollow without a top, then there will be one area of a circle to consider.

The volume of the cylinder is given by

$$V = \pi r^2 h$$

EXAMPLE 9

Find the area of metal in a metal can of radius 3.3 cm and height 10 cm correct to 1 place of decimals. Calculate too, the volume of the can.

$$\begin{aligned}\text{Area of can around side} &= 2\pi rh \\ &= 2 \times 3.142 \times 3.3 \times 10 \\ &= 207.37 \text{ cm}^2\end{aligned}$$

$$\begin{aligned}\text{Area of top} &= \pi r^2 \\ &= 3.142 \times 3.3 \times 3.3 \\ &= 34.22 \text{ cm}^2\end{aligned}$$

$$\text{Area of bottom} = 34.22 \text{ cm}^2$$

$$\begin{aligned}\text{Total area} &= 275.81 \\ &= 275.8 \text{ cm}^2 \quad \text{to 1 place of decimals}\end{aligned}$$

$$\begin{aligned}\text{Volume of can} &= \pi r^2 h \\ &= 3.142 \times 3.3 \times 3.3 \times 10 \\ &= 342.2 \text{ ml}\end{aligned}$$

(d) *THE SPHERE*

The sphere or globe has the following formulae:

$$\text{Surface area} = 4\pi r^2$$

$$\text{Volume} = \tfrac{4}{3}\pi r^3$$

EXAMPLE 10

Find the amount of sheet metal in a hollow aluminium globe of radius 15 cm. Calculate its volume.

$$\begin{aligned}\text{Surface area} &= 4\pi r^2 \\ &= 4\pi \times 15 \times 15 \\ &= 2827.8 \text{ cm}^2\end{aligned}$$

$$\begin{aligned}\text{Volume} &= \tfrac{4}{3}\pi r^3 \\ &= \tfrac{4}{3} \times 3.142 \times 15 \times 15 \times 15 \\ &= 14{,}137 \text{ ml}\end{aligned}$$

(e) *THE CONE*

The cone is shown in Fig. 8.16. Its volume is given by the formula:

$$V = \frac{1}{3}\pi r^2 h$$

The volume of a cone is seen to be $\frac{1}{3}$ that of a cylinder with the same height and same base radius.

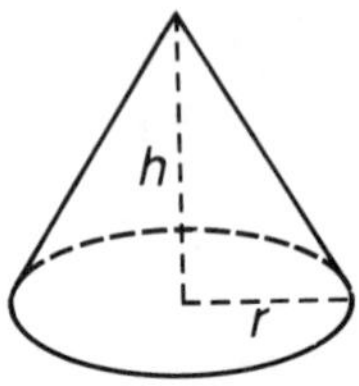

FIG. 8.16 The cone

EXERCISES 8.3

1. Find the volume and surface area of the following:
   (a) Solid cylinder, base radius 7 cm, height 16 cm
   (b) Sphere, radius 8 cm
   (c) Hollow cylinder, radius 5 cm, height 8 cm, no top
   (d) Sphere, radius 6.2 cm.

2. Find the volume of a cone, base radius 6 cm, height 1.1 cm.

3. Find the volume of a three-tier, circular cake, each tier having a depth of 11 cm, the tiers have radii of 34 cm, 24 cm, 16 cm.

4. A drum has a radius of 30 cm and height 75 cm. It contains vinegar. How many bottles of volume 75 cl can be filled from it?

5. Calculate the area of a sheet of metal required to make the following cans:
   (i) Tongue: height 5 cm, radius 3.7 cm
   (ii) Tinned peas: height 10 cm, radius 3.5 cm
   (iii) Baked beans: height 11.5 cm, radius 3.7 cm
   (iv) Sliced pineapple: height 9 cm, radius 4.7 cm.

6. A circular garden, radius 20 m, is completely surrounded by a path 1.5 m wide. Find the total area of the path.

7. Find the area of each of the shaded portions shown in Fig. 8.17.

MISCELLANEOUS EXERCISES 8

1. The diagram in Fig. 8.18 shows a floor plan of a hotel restaurant. Calculate the number of square metres of carpet required to cover the floor completely. Calculate the cost if the carpet is £12.50 per m^2.

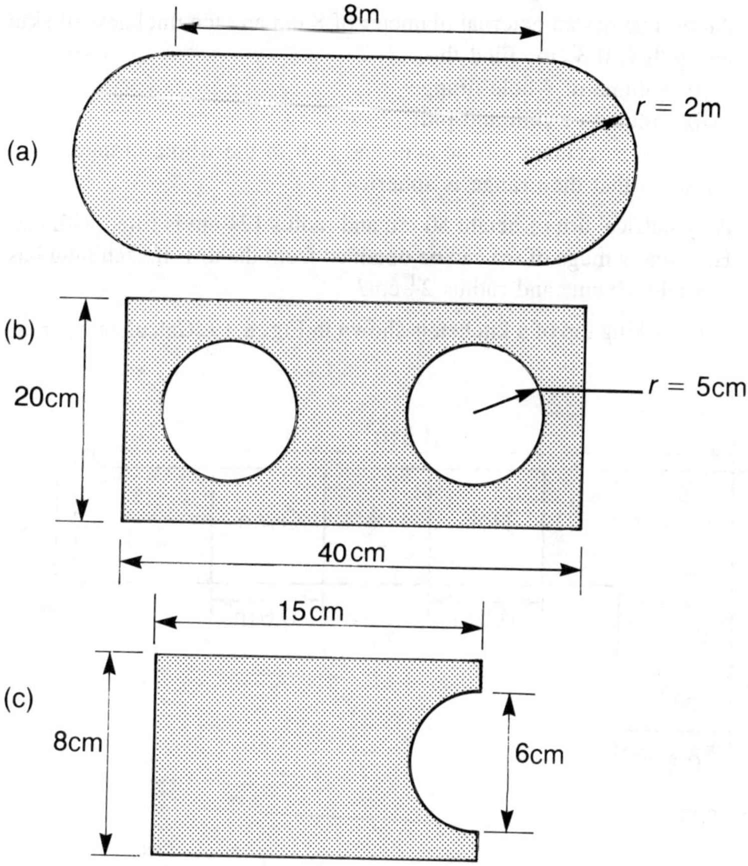

FIG. 8.17

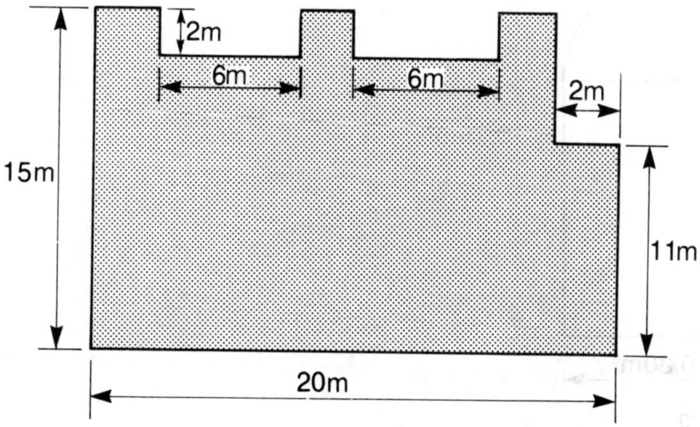

FIG. 8.18

2. An orange has an external diameter of 8 cm and the thickness of skin and pith is 0.3 cm. Find the:
(i) volume of whole orange
(ii) volume of skin and pith
(iii) the percentage of the skin and pith in the whole orange.
(Assume that the orange is spherical).

3. A cylindrical urn of height 50 cm and radius $12\frac{1}{2}$ cm is filled with tea. How many mugs of tea can be obtained from the urn, if each mug has a height 10 cm, and radius $2\frac{1}{2}$ cm?

4. The working top of a kitchen is shown in Fig. 8.19. Calculate its area.

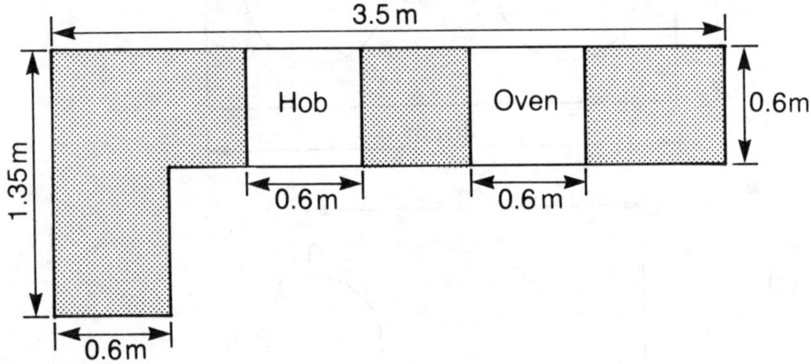

FIG. 8.19

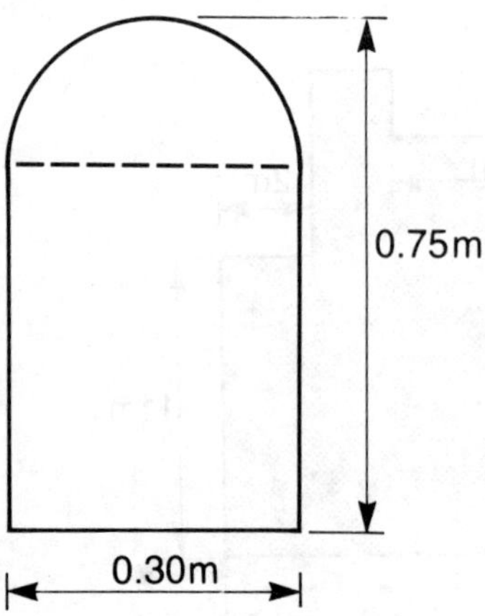

FIG. 8.20

5. Figure 8.20 shows a diagram of a hot-water tank, which comprises of a cylinder, with a hemisphere on top. The dimensions are radius 0.15 m, total height 0.75 m. Calculate the total area of copper sheet in the tank and its volume.

6. A bucket is shown in Fig. 8.21. It can be thought of as the bottom part of a cone. Calculate its volume.

7. A wall of a bathroom has dimenions 12 m × 3.5 m. Find the number of tiles 15 cm × 15 cm needed to cover the wall completely.

8. A wine bottle has a volume of 70 cl. Its radius at the base is 3.5 cm. Find the height of the bottle up to the neck.

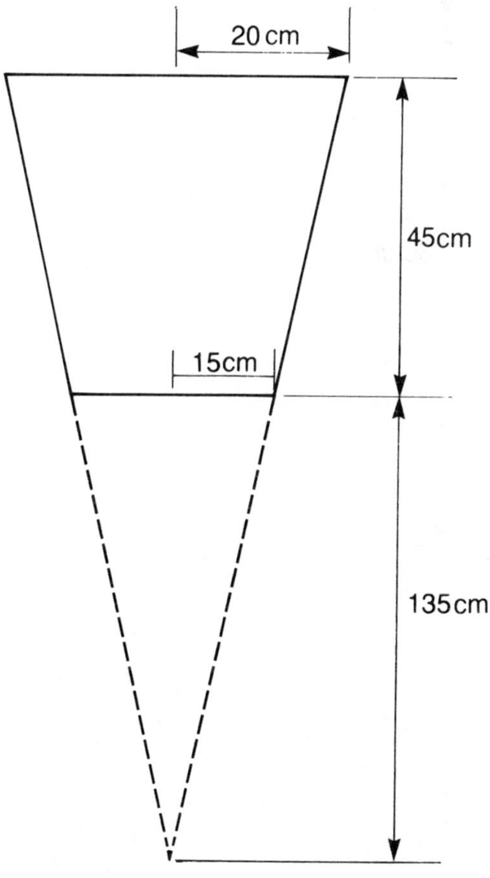

FIG. 8.21

9. A wine glass can be thought of as the bottom of a cone, as shown in Fig. 8.22. Wine is poured into the glass up to 2 cm from the rim. Find the volume of wine in the glass. Calculate how many glasses may be obtained from a 75 cl bottle.

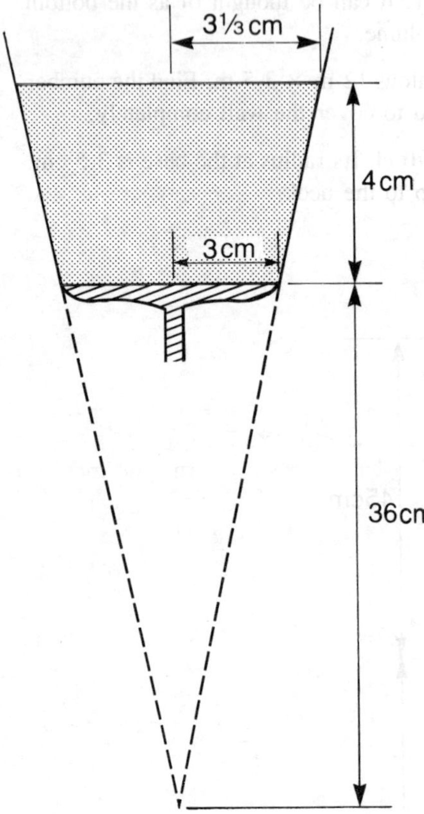

FIG. 8.22

9

DATA HANDLING AND PRESENTATION

The branch of mathematics concerned with data handling is called **statistics**. It is concerned with counting and measuring, handling large numbers of figures, classifying them to obtain patterns, and presenting them in a way that can readily be understood.

For example, we could have tables of the height of all adults in the country. Such a list would be unmanageable and have little worth. On the other hand, such tables could be averaged so that fewer numbers could be considered, for example, the separate average heights of men and women. Then a comparison could be made between the heights of men and women.

Data can be collected by measuring, for example, the diameter of cylinders produced in manufacturing, or by counting, such as the daily occupation of rooms in a hotel over a given period. Having obtained the figures, it is then necessary to classify them in order to detect patterns and trends.

One may ask how the diameters of cylinders vary if each is supposed to have a set value. Whether control is by machine or man, variations occur at random and the measurements vary about the set value. For example, if the cyclinder were to have a diameter of 35 mm, the actual diameters of a series might look like this:

35.3, 34.4, 33.7, 36.2, etc.

9.1 CLASSIFICATION

Figures that are obtained by measuring or counting must be classified into groups in order to make the data manageable. The method of doing this is illustrated using the following example.

The figures shown in Table 1 are the daily units of electricity used in a restaurant over 60 days. In recording the readings the meter dial figures are corrected to the nearest unit. Thus 148.4 kW h becomes 148 kW h and 161.7 kW h becomes 162 kW h. These readings are as shown in Table 1.

(a) *RANGE*

From Table 1 we see that the maximum number of units used is 187 and the lowest 131. The difference between maximum reading and minimum reading is called the **range**. In our example, therefore

range = 187 − 131 = 56 units

131	162	155	166	152	147	184	157	182	154
157	142	166	151	167	173	154	146	179	137
144	187	172	135	155	149	164	162	158	165
151	158	159	156	175	165	162	153	143	155
179	167	155	161	166	145	176	161	167	169
153	148	142	168	159	156	163	161	151	155

TABLE 1

(b) *CLASS INTERVALS*

The range is then divided up into a number of intervals which may or may not be equal. It is more convenient to have equal intervals and we shall adopt this in our work. These are called **classes** or **class intervals**.

As a rough rule, the number of intervals should be between 5 and 15 so that a pattern can be discernible. In our case with the range of 56 we can make the class interval 10, thus giving 6 intervals altogether.

The first class interval will be 130 to 139. However, the real boundaries are 129.5 and <139.5, because all the measurements had been corrected to the nearest unit. The <139.5 means up to 139.5, because 139.5 goes into the next class. The class intervals are therefore:

129.5–<139.5
139.5–<149.5
149.5–<159.5
159.5–<169.5
169.5–<179.5
179.5–<189.5

(c) *FREQUENCY*

The next part of the classification exercise is to find from the original list of units how many times the daily readings fall into each class. For example, looking through Table 1, the number of days in which the electricity consumption falls between 129.5 and <139.5 is 3. This number 3 is called the **frequency**. This is carried out until all the figures in Table 1 have been listed into one or other of the classes.

(d) *TALLY DIAGRAM*

From Table 1 we take the first reading, 131. This will be in the first class and so we place a 1 to represent this reading. We then move on to the second,

162, which is in the class 159.5 to 169.5. A tally of 1 is added to the appropriate class for each reading, grouping them each time in fives, 𝍸. The complete table is shown in Table 2.

Class interval	Tally	Class frequency
129.5–<139.5	III	3
139.5–<149.5	𝍸 IIII	9
149.5–<159.5	𝍸 𝍸 𝍸 𝍸 I	21
159.5–<169.5	𝍸 𝍸 𝍸 III	18
169.5–<179.5	𝍸 I	6
179.5–<189.5	III	3

TABLE 2

(e) *RELATIVE FREQUENCY AND PERCENTAGE RELATIVE FREQUENCY*

Relative frequency is $\dfrac{\text{class frequency}}{\text{total}}$.

Thus, the relative frequencies of the above data are $\frac{3}{60}$, $\frac{9}{60}$, $\frac{21}{60}$, etc. The percentage relative frequency is then each of these figures × 100. The complete table is shown in Table 3.

Class interval	Class frequency	Relative frequency	% Relative frequency
129.5–<139.5	3	$\frac{1}{20}$	5
139.5–<149.5	9	$\frac{3}{20}$	15
149.5–<159.5	21	$\frac{7}{20}$	35
159.5–<169.5	18	$\frac{6}{20}$	30
169.5–<179.5	6	$\frac{2}{20}$	10
179.5–<189.5	3	$\frac{1}{20}$	5

TABLE 3

9.2 REPRESENTATION OF CLASS FREQUENCY

(a) *BAR CHART*

The class frequencies in Table 2 can be represented by bar charts, drawn either vertically or horizontally. The class frequency is represented by a bar whose length is drawn to scale.

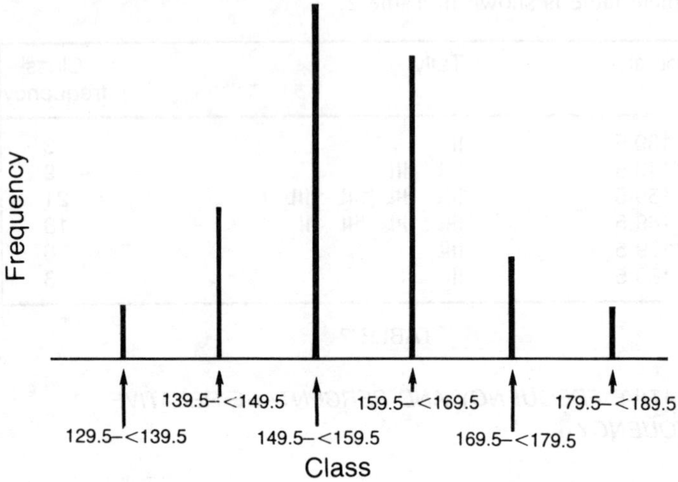

FIG. 9.1 Bar chart

(b) *PICTOGRAM*

This is an eye-catching method of showing the variation of class frequencies. A small picture, called a **pictogram**, is drawn to represent a convenient number of class frequencies. In our example, let the pictogram

 ≡ 3 class frequencies

Hence, we can represent the first class by 1 pictogram, the second by 3 pictograms, and so on. The final result is shown in Fig. 9.2.

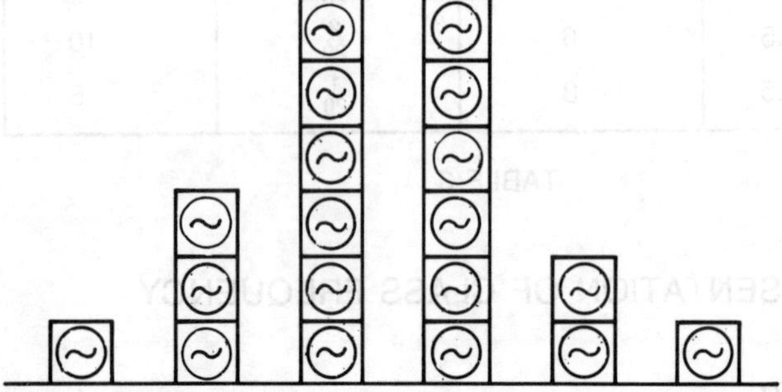

FIG. 9.2 Pictogram

9.3 PICTORIAL REPRESENTATION OF THE RELATIVE FREQUENCY

The classification in Table 3 can now be shown pictorially using (a) a 100% bar chart, or (b) a pie diagram or pie chart.

(a) *BAR CHART*

The percentage relative frequency is shown to scale on a bar chart, as shown in Fig. 9.3.

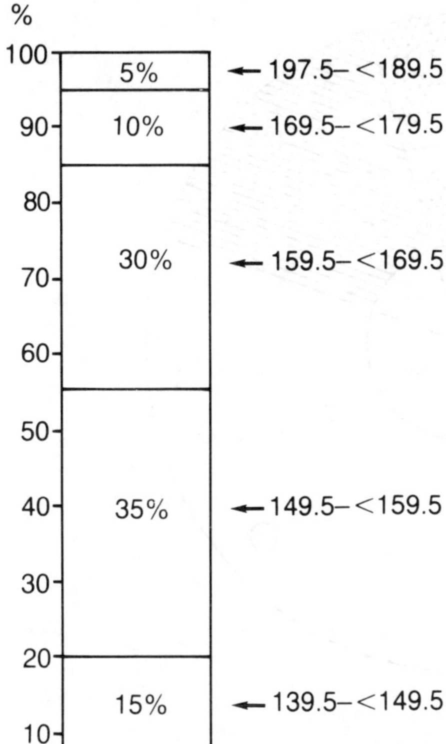

FIG. 9.3 The 100% bar chart

(b) *PIE CHART*

In the pie chart, the total angle at the centre of the circle is 360°. Each class frequency is allocated a proportion of 360° according to the percentage relative frequency. In the example the angle allocated to the:

first class	=	5% of 360°	=	18°
second class	=	15% of 360°	=	54°
third class	=	35% of 360°	=	126°
fourth class	=	30% of 360°	=	108°
fifth class	=	10% of 360°	=	36°
sixth class	=	5% of 360°	=	18°

The pie chart is shown in Fig. 9.4, where each class is represented by a sector, with the angles listed above.

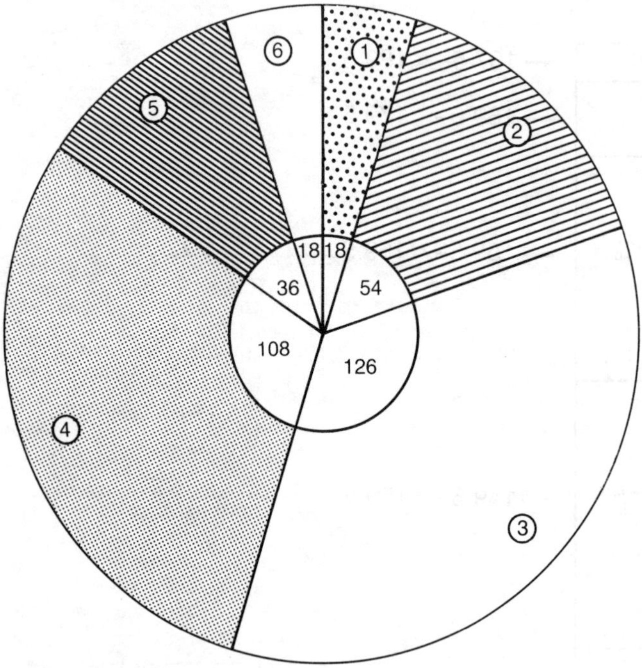

FIG. 9.4 Pie chart of the data

EXAMPLE 1

The capacity of 20 wine bottles, all nominally 3 ℓ in volume, were measured to the nearest 0.1 ℓ. The results were as follows:

2.8	3.0	2.9	3.0	3.0	2.9	3.1
3.1	2.8	3.0	3.2	2.9	2.8	2.9
3.0	2.9	2.9	2.9	3.1	3.2	

Summarise these data with a tally chart, obtain a table for class frequency and draw a bar chart to depict the analysis.

(Neath C)

Tally diagram

Volume		Frequency
2.8	III	3
2.9	~~IIII~~ II	7
3.0	~~IIII~~	5
3.1	III	3
3.2	II	2

Bar chart

We can use a scale of 1 frequency represented by 1 cm, so that the bar chart is shown in Fig. 9.5.

It is worth noting that each of the various classes representing the volume of the bottles to the nearest 0.1 ℓ, has a class interval of 0.1 ℓ, that is, the first class represents volumes of bottles from 2.75 up to 2.85, the second class from 2.85 and up to 2.95, and so on.

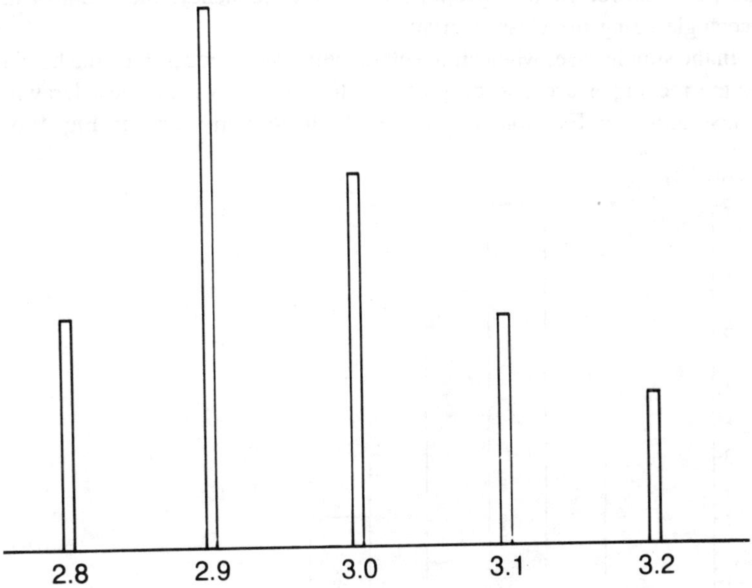

FIG. 9.5 Bar chart for volumes of wine bottles

9.4 CONTINUOUS AND DISCRETE DATA

In our work so far the data collected and analysed have been continuous, in that within a class interval any value is possible. For example, in measuring electrical energy units, within the first class all values between 129.5 and 139.5 kW h are possible, and so on. Such data, typical of all measurements, are **continuous**.

On the other hand, data may be collected by counting. For example, the number of matches in a box, or the number of rooms occupied in a hotel over a given period. Such counts produce **discrete** data, in that they can only be whole numbers. The nature of discrete data does not affect the methods of presentation already described in sections 9.2 and 9.3.

However, in the next section we will explain what is meant by a histogram. In this case the nature of the data, whether it is continuous or discrete, becomes important.

9.5 HISTOGRAM

Simply, the frequency of each class is plotted as rectangles such that the *area* of each rectangle is proportional to the frequency, the width of the rectangle being the class interval.

In the simple case, when all the class intervals are equal, then the heights of the rectangles are also proportional to the class frequencies. Drawing a histogram for Example 1 produces the histogram shown in Fig. 9.6.

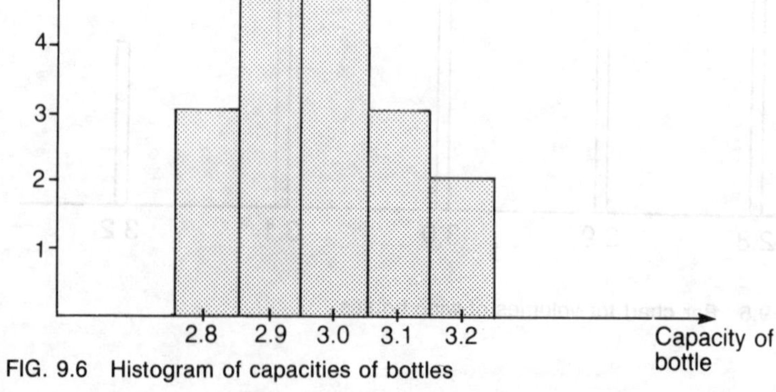

FIG. 9.6 Histogram of capacities of bottles

The overriding requirement still holds, that is, the frequency is proportional to the area of each rectangle.

When classes are not equal, the height of each rectangle is obtained by making the area proportional to the frequency, as shown in Example 2.

EXAMPLE 2

Fifty students on a catering course were asked to measure out a quantity of flour and having done so, to weigh each quantity. They found the following distribution:

No. of students (frequency)	3	6	28	11	2
Weights of flour in g (class)	245–<247	247–<249	249–<253 *	253–<255	255–<257

Plot a histogram of the data.

The measurements of weights produce continuous data, since any value within each class is possible. The class intervals are equal except for the middle one, marked by *.

A histogram is such that the area of each rectangle is proportional to the frequency. Because the class, with interval 249–<253, is twice the others, the height of the bar must be half that appropriate to a class frequency of 28. If it had been three times the width, then the bar height would be one-third the height proportional to that particular class frequency.

The reasoning is as follows: 28 students had measured weights in the range 249–<253 and it is assumed that the individual weights are spread fairly evenly over the class interval. Hence, only 14 would be expected to have weights of flour in the range 249–<251 and another 14 in the range 251–<253. Hence, the bar height corresponds to a class frequency of 14. This is shown by the shaded area in Fig. 9.7.

The latter analysis is true for all distributions having variable class intervals. The problem is thus overcome by making

area of each rectangle proportional to class frequency

The histogram of the data in Example 2 is shown in Fig. 9.7.

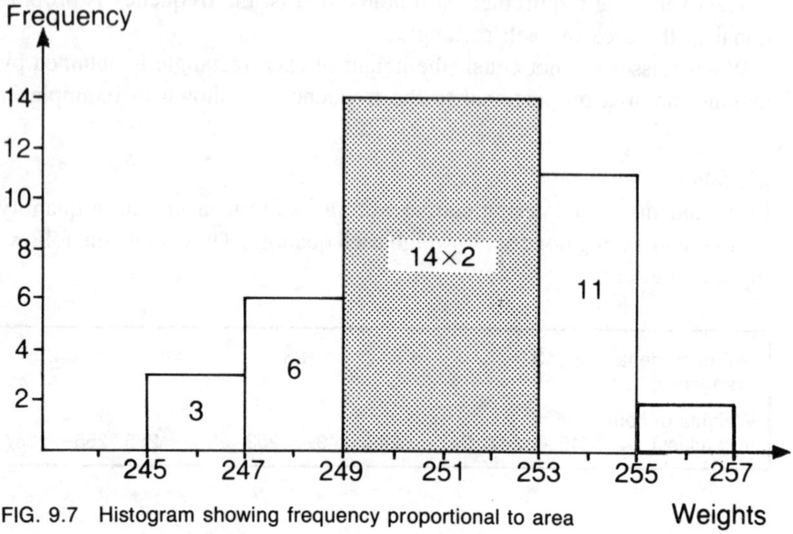

FIG. 9.7 Histogram showing frequency proportional to area

Note: A histogram may be plotted with relative frequencies and percentage relative frequencies instead of actual class frequencies. In this case the total relative frequencies add up to 1 and the percentage relative frequencies add up to 100.

In the case where a histogram is required for discrete data, the class boundaries do not exist. For convenience, however, rectangles are drawn to touch one another as for the continuous data.

EXERCISES 9.1

1. Which of the following are discrete variables:
(a) diameter of apples
(b) number of cups of tea sold in the canteen per day
(c) daily kitchen temperature
(d) number of phone calls made per day
(e) lifetime of electric lightbulbs.

(Neath C)

2. The table below shows the people employed in a factory canteen:

unskilled workers	45
ancillary catering staff	25
cooks	5
clerical staff	10

Draw a bar chart to represent the data. In order to provide a pictorial display, draw a pie chart.

(Neath C)

3. The examination marks for 50 students were distributed as follows:

Marks	0–<20	20–<30	30–<40	40–<50	50–<60	60–<70	70–<100
No. of students	4	5	6	15	6	5	9

Plot a histogram of the distribution.

4. In a survey of 240 people who had taken 'GCSE' level mathematics:
120 passed at the first attempt
60 passed at the second attempt
30 passed at the third attempt
20 passed at the fourth attempt
10 passed at the fifth attempt.
Show this information in a pie diagram. (Nelson and Colne)

5. Fifteen 1983 10p coins, all completely new, were weighed separately in grams. Their weights were:
12.65, 12.66, 12.80, 12.80, 12.48, 12.71, 12.73, 12.73, 12.78, 12.64, 12.82, 12.63, 12.80, 12.77, 12.71.
Plot a histogram of the distribution of weight. (Nelson and Colne)

6. The following table shows the number of hours of rainfall per day at a seaside resort during the month of April.

9.2	0.8	6.1	9.9	5.3
3.5	5.8	6.4	1.2	2.1
1.1	0.8	5.2	8.4	3.7
4.6	5.3	8.1	6.3	3.9
8.7	5.1	0.5	1.8	4.2
6.3	0.7	1.4	3.5	7.1

Copy and complete the following tally and frequency tabulation:

Class	Tally	Frequency
0–>2		
2–>4		
4–>6		
6–>8		
8–>10		

Represent the above frequency distribution by a histogram (on graph paper).

(Nelson and Colne)

7. The following figures are the total costs for a restaurant in respect of half a year's turnover:

Cost of sales:	£
Food costs	24.000
Beverage costs	7,000
Other costs	9,000
Labour costs:	£
Wages	19,000
National Insurance	2,500
Staff meals	2,000
Staff welfare	500
Overheads:	
Rent and rates	6,000
Depreciation	4,000
Repairs	1,000
Light and heat	3,000
Administration	5,000
Other expenses	1,000

(a) Use a pie diagram to represent data under 3 general headings:
Sales
Labour
Overheads

(b) Use a vertical bar chart to show data for individual expenses in overheads section.

(Oxford)

8. Specify the data type (discrete or continuous) for each of the following variables:

(a) average daily rainfall (cm)
(b) number of rooms in a hotel
(c) weight of a lady guest (kg)
(d) collar sizes (cm)
(e) number of children per family.

9. Classify the following data as discrete or continuous:

(a) the number of meals served in a restaurant each day
(b) the cost of a meal
(c) the number of people served
(d) the time spent waiting to give the order
(e) the weight of meat in the meal
(f) the lifetime of an electric light bulb

(g) the number of bedrooms in a hotel.

Add to the list four more variables, stating whether they are discrete or continuous.

(GLOSCAT)

10. A spot check at two bars in a hotel revealed the following numbers of people drinking certain drinks. Draw two pictograms to show the two distributions and comment briefly on the figures.

	Bar 1	Bar 2
Glass of wine	3	9
Pint of beer/lager, etc.	10	2
Glass of spirits	2	5
Soft drinks	3	4

(GLOSCAT)

11. Construct a pie chart and a percentage component bar chart to represent the following data concerning guests at a hotel. The guests were classified according to the area in which they normally live.

Area	No. of guests
UK	37
EEC (excl. UK)	11
USA	15
Commonwealth	9
Other areas	3

(GLOSCAT)

12. The hotel bills of 40 guests over a Christmas period are shown below. Arrange the amounts in groups and form a frequency distribution.
70, 96, 114, 98, 126, 56, 52, 104, 106, 95, 80, 75, 76, 78, 145, 110, 160, 84, 163, 127, 168, 141, 112, 127, 122, 87, 119, 106, 150, 109, 154, 58, 92, 78, 98, 95, 100, 126, 90, 118.

(Blackpool and Fylde C of HE)

13. The following is a record of marks scored by candidates in an examination:

77	59	84	73	51	43	50	81	61	53	69
37	58	63	67	61	90	61	50	60	84	56
77	57	42	43	41	49	37	21	24	35	34
50	11	52	30	16	33	67	87	64	47	59
37	92	88	30	38	22	22	49	46	50	64
23	73	73	48	26	36	51	85	71	57	45

(a) Tabulate the marks in the form of a frequency distribution using a tally chart, grouping by suitable intervals.
(b) Construct a histogram from your frequency distribution.
(c) Explain the essential differences between a histogram and a bar chart.

(Blackpool and Fylde)

ANSWERS TO EXERCISES

CHAPTER 1

EXERCISES 1.1

1. 6 **2.** 2 **3.** 4 **4.** 5 **5.** 17
6. 1 **7.** 6 **8.** 7 **9.** 13 **10.** 27

EXERCISES 1.2

1. (a) $\frac{4}{5}$ (b) $\frac{3}{4}$ (c) $\frac{3}{5}$ (d) $\frac{1}{3}$ (e) $\frac{3}{13}$

2. (a) $8\frac{1}{2}$ (b) $1\frac{2}{7}$ (c) $12\frac{4}{9}$ (d) $1\frac{1}{13}$

3. (a) $\frac{12}{7}$ (b) $\frac{13}{3}$ (c) $\frac{12}{5}$ (d) $\frac{43}{7}$

EXERCISES 1.3

1. (a) $\frac{19}{20}$ (b) $1\frac{13}{24}$ (c) $1\frac{3}{20}$ (d) $9\frac{9}{28}$ (e) $5\frac{25}{36}$ (f) $9\frac{7}{8}$

2. (a) $\frac{5}{21}$ (b) $3\frac{1}{36}$ (c) $2\frac{7}{20}$ (d) $1\frac{5}{6}$ (e) $2\frac{17}{24}$ (f) $2\frac{10}{21}$

3. (a) $4\frac{3}{4}$ (b) $1\frac{7}{8}$ (c) $3\frac{1}{6}$ (d) $4\frac{11}{28}$ (e) $2\frac{17}{42}$ (f) $5\frac{1}{10}$

EXERCISES 1.4

1. 15 **2.** $1\frac{5}{21}$ **3.** $13\frac{1}{3}$ **4.** $9\frac{1}{6}$ **5.** $\frac{5}{6}$

6. $2\frac{1}{12}$ **7.** $1\frac{3}{5}$ **8.** $1\frac{4}{5}$ **9.** 11 **10.** $6\frac{3}{4}$

11. $\frac{3}{10}$ **12.** 1

EXERCISES 1.5

1. $3\frac{1}{2}$ **2.** $3\frac{11}{16}$ **3.** $\frac{2}{9}$ **4.** $1\frac{19}{24}$ **5.** $2\frac{9}{14}$

6. $3\frac{7}{9}$ **7.** $4\frac{1}{20}$ **8.** $13\frac{1}{2}$

MISCELLANEOUS EXERCISES 1

1. (a) $\frac{1}{3}$ (b) $\frac{13}{16}$ (c) $\frac{3}{5}$ (d) $4\frac{2}{3}$ (e) $\frac{2}{3}$ (f) $12\frac{5}{6}$

2. $62\frac{13}{20}$ **3.** $7\frac{7}{12}$ **4.** $7\frac{1}{6}$ **5.** $18\frac{11}{20}$, $15\frac{5}{12}$, $3\frac{2}{15}$
6. $4\frac{1}{3}$ m^2, $3\frac{2}{3}$ m^2 **7.** £$11\frac{2}{5}$ **8.** $15\frac{3}{4}$ hours **9.** 9
10. 13 days **11.** (a) $15\frac{1}{2}$, (b) $15\frac{1}{2}$ m^2 **12.** 16 **13.** $\frac{29}{32}$ of a bottle

CHAPTER 2

EXERCISES 2.1

1. (a) $\frac{1}{2}$ (b) $\frac{3}{5}$ (c) $\frac{9}{20}$ (d) $\frac{16}{25}$ (e) $\frac{18}{25}$
(f) $\frac{24}{25}$ (g) $\frac{29}{40}$ (h) $1\frac{1}{2}$ (i) $2\frac{1}{4}$ (j) $3\frac{11}{50}$

2. (a) 0.4 (b) 0.34 (c) 0.16 (d) 0.04
(e) 0.031 (f) 0.006 (g) 0.25 (h) 0.2
(i) 0.8 (j) 0.875 (k) 0.625

EXERCISES 2.2

1. (a) 17.66 (b) 5.87 (c) 18.96 (d) 19.94
(e) 0.08 (f) 3.01 (g) 4.00 (h) 8.96
(i) 12.00 (j) 7.78

2. (a) 17.7 (b) 4.32 (c) 5.82 (d) 7.94
(e) 176 (f) 184 (g) 1850 (h) 19 700
(i) 0.00812 (j) 0.0664 (k) 9.06 (l) 3.01

3. (a) 36.3932 (b) 10.004 (c) 24.427
(d) 0.729041 (e) 302.602 (f) 20.0928
(g) 6.71 (h) 7.02

MISCELLANEOUS EXERCISES 2

1. (a) 0.375 (b) 0.273 (c) 0.4 (d) 0.571
(e) 0.231 (f) 0.857 (g) 0.875
2. 2.67
3. (a) 0.18 (b) 0.0176 (c) 0.5 (d) 3.06 (e) 3.0
4. (a) 24.677 (b) 6.64 (c) 13.34 (d) 20.40

5. (a) $\frac{4}{25}$ (b) $\frac{6}{25}$ (c) $\frac{11}{50}$ (d) $\frac{9}{50}$ (e) $\frac{23}{25}$ (f) $\frac{5}{8}$

CHAPTER 3

EXERCISES 3.1

1. $\frac{7}{20}$, 0.35; $\frac{7}{25}$, 0.28; $\frac{2}{5}$, 0.04; $\frac{5}{8}$, 0.625
$\frac{9}{10}$, 0.9; $\frac{2}{3}$, 0.667; $\frac{3}{25}$, 0.12; $\frac{17}{100}$, 0.17
$\frac{31}{100}$, 0.31

2. 37.5, 60, 90, 55, 16, 61, 55,
81.6, 79, 0.5

EXERCISES 3.2

1. (a) £80 (b) 210 m^2 (c) £6.75 (d) 6.6 kg (e) 10.15 m^2
2. 4800 m^2 **3.** 67.5 m^2 **4.** 18, £148.50
5. £452.40 **6.** 1.92 kg **7.** £368 **8.** 27
9. 48.75 kg, 390 **10.** 3960 units **11.** £10 935
12. 0.36 kg, 5.16 kg **13.** £114

EXERCISES 3.3

1. (a) 10% (b) 4% (c) 88.9%
2. 4% **3.** 16.7%, 60%, 15.6%, 7.8%
4. 10% **5.** (i) 2.1% (ii) 87.8% (iii) 40%
6. 52.2% — second month
7.

	Cheddar	Cottage
Protein	26	14
Fat	34	4
Carbohydrate	0	1
Mineral	3	1
Water	37	80

8. 7.7%, 10.8% **9.** 25%, 37.5%, 12.5%, 6.25%, 6.25%, 12.5%
10. 24.5%, 38.25%, 8.25%, 29%

EXERCISES 3.4

1. £160 **2.** 200 m^2 **3.** 200 boxes **4.** £3072
5. £3200 **6.** 88 litres **7.** £1000, 87p **8.** 1500
9. 425 **10.** 48 hrs **11.** 40 weeks

MISCELLANEOUS EXERCISES 3

1. (a) $\frac{13}{20}$ (b) $\frac{9}{20}$ (c) $\frac{1}{8}$ (d) $\frac{3}{50}$
2. (a) 600 kg (b) £72 (c) 165 m^2 (d) 390 km
3. (a) £3200 (b) 900 ft^2 (c) 3000 miles (d) 540 boxes
4. (a) 75% (b) 16.7% (c) 10% (d) 6.25%
5. (a) 812.7 (b) £429 (c) 477 kg (d) 54°C
6. (a) 1.08% (b) 0.52% (c) 0.27; hence File C is suitable
7. £338, £422.50, 62.5%
8. 221 g, 221 g, 331.5 g, 42.5 g, 25.5 g, 8.5 g
9. £10 935
10. £1295

CHAPTER 4

EXERCISES 4.1

1. (a) 60%, (b) 60.4%, (c) 72.3%
2. (a) 43.3% (b) 44.4% (c) 46.8%
3. £5.10, 85%
4. (a) £36, 60%, (b) £60, 60%, (c) £60, 50%
5. 75%
6. 60% **7.** (i) 70% (ii) 20%
8. (i) £3150, 900%, (ii) £335, 10.6%
9. 30.8%
10. (i) 75% (ii) 41.9%
11. 86%, 74%, 68%

EXERCISES 4.2

1. £391 **2.** £51.75 **3.** £862.50 **4.** 39p
5. £5.87
6.

<table>
<tr><td>£4.03</td><td>£138</td><td>£724.50</td><td>£16.09</td><td>£150</td></tr>
</table>

7. £63.80, £8.32
8. £56.00

EXERCISES 4.3

1. (a) £32 (b) £6250 (c) £1830
2. (a) £35 (b) £20
3. £5000
4. £1125

MISCELLANEOUS EXERCISES 4

1. £53.36 **2.** 8% **3.** £210 **4.** £8.35 **5.** £5500 **6.** £13 000
7. (a) 53.6, 46.8, 43.5, (b) 43.4% (c) 48.1% **8.** £6000 **9.** 2 lb, 80p
10. 42.7%

CHAPTER 5

EXERCISES 5.1

1. 6.3 **2.** 10.17 **3.** 15.98 **4.** 3.43
5. 76.00 **6.** 0.63 **7.** 4.10 **8.** 13.40

EXERCISES 5.2

1. 25.30 **2.** 1.28 **3.** 23.79 **4.** 10.78
5. 10.84

EXERCISES 5.3

1. 34.363 **2.** 4.306 **3.** 14.749 **4.** 11.856

EXERCISES 5.4

1. 30 **2.** 4 **3.** 2 **4.** 1 **5.** 2

CHAPTER 7

EXERCISES 7.1

1. (i) 96 (ii) 64

2. (i) 140 (ii) 93 (iii) 70
3. 28
4. 90
5. 12.5%, 156
6. 25%, 25
7. 23 lb, 61
8. 3 kg, $3\frac{1}{3}$ kg

EXERCISES 7.2

1. £75, £4.69, £9.38
2. (i) £36 (ii) 33 (iii) £1.09
3. £160, 160, £1
4. £125, 75, £1.67
5. 95p, £2.11
6. $15\frac{3}{4}$ lb, $5\frac{1}{4}$ oz, 83p

EXERCISES 7.3

1. 69p, £1.38 **2.** 40p, £1.33
3. £16.50, £41.25

EXERCISES 7.4

1. £13.30 **2.** £271.59 **3.** 59.8%
4. £32.20, £16.65, £7.20 **5.** £25.60, 80p
6. 45p

EXERCISES 7.5

1. £45.50 **2.** £24.19 **3.** Total due £108.80
4. Total due £63.26 **5.** 12 gallons, £21.60, £2.82

MISCELLANEOUS EXERCISES 7

1. (i) 600 g (ii) 42p (iii) £1.40
2. Total account: £211.30
3. Total account: £123.62
4. (a) 56 lb (b) 91 portions (c) 28p (d) 32% (e) 72p per portion
5. 40p, 53p, 47p
6. £648.55

CHAPTER 8

(*Note*: $cm^3 = ml$)

EXERCISES 8.1

1. (a) $16\ m^2$ (b) $36\ m^2$ (c) $91\ m^2$ (d) $32\ m^2$
(e) $15\frac{2}{5}\ m^2$ (f) $32.1\ m^2$ (g) $59.8\ m^2$

2. (a) $32\ m^2$ (b) $76\ m^2$ (c) $106\ m^2$ (d) $31\ m^2$

3. $39\ m^2$, 5

4. $85.6\ m^2$

5. $47.5\ m^2$, £283.88, £37.03

EXERCISES 8.2

1. (a) 0.45 ℓ (b) 0.216 ℓ (c) 0.594 ℓ
(d) 0.6015 ℓ (e) 0.6985 ℓ

2. $56.25\ m^3$, $11\frac{1}{4}$ min.

EXERCISES 8.3

1. (a) $2463\ cm^3$, $1012\ cm^2$ (b) $2145\ cm^3$, $804\ cm^2$
(c) $628\ cm^3$, $300\ cm^2$ (d) $317\ cm^3$, $483\ cm^2$

2. $41.5\ cm^3$ **3.** $68\,700\ cm^3$ **4.** 212 ℓ, 282 bottles

5. (i) $202\ cm^2$, $215\ cm^3$ (ii) $297\ cm^2$, $385\ cm^3$
(iii) $353\ cm^2$, $495\ cm^3$ (iv) $405\ cm^2$, $625\ cm^3$

6. $196\ m^2$

7. (a) $44.6\ m^2$ (b) $643\ cm^2$ (c) $106\ cm^2$

MISCELLANEOUS EXERCISES 8

1. $268\ m^2$, £3350

2. $268\ cm^3$, $56\ cm^3$, 20.9%

3. 125 cups

4. $1.83\ m^2$

5. $0.78\ m^2$, $0.05\ m^3$

6. 43.6 l

7. 1867 tiles

8. 18.2 cm

9. 126 ml, 6 glasses

INDEX